Chickadee Tales

A New Haven Bird Club Anthology

Edited by
Gail Martino and Ricci Cummings

Copyright © 2021 by New Haven Bird Club, Inc.

All rights reserved. No part of this book may be reproduced or used in any manner without written permission of the copyright owner except for the use of quotations in a book review. For more information, address: ChickadeeTales@newhavenbirdclub.org

First paperback edition January 2021

Book design by Hannah Gaskamp

Cover design by Hannah Gaskamp

ISBN 978-1-7357122-9-1 (paperback)

ISBN 978-1-7357122-8-4 (e-book)

www.NewHavenBirdClub.org

*This book is dedicated to all New Haven Bird Club members,
past, present, and future.*

Contents

PREFACE 1

CHICKADEE TALES: AN INTRODUCTION 5

PART I
Foundations 11

1 A Century and More 13
   Craig Repasz

PART II
Birder Beginnings 51

2 Why We Bird 53
   Gail Martino

3 Interview with Arne Rosengren 65
   Florence McBride and DeWitt Allen

PART III
Special Programs 69

4 Ed's Count 71
   Frank Gallo

5 Hawk Watch 75
   Steve Mayo

6	Mega Bowl *Chris Loscalzo*	81
7	Bird Walks and Those Who Lead Them *Ricci Cummings*	89

PART IV
Education and Conservation 97

8	*Take Flight!* Bird Observation with Children in a School Science Program and Beyond *Florence McBride*	99
9	Restoring Local Populations of American Kestrels, One Box at a Time *Tom Sayers*	111
10	A Condo Development That Bluebirds Love *Pat Leahy*	121

PART V
Tributes 127

11	There's Always Something to See: A Tribute to Noble Proctor *Dan Barvir, Frank Gallo, Patrick Lynch, and Florence McBride*	129
12	The Gifts of Richard English *Martha Lee Asarisi and Michael Horn*	137
13	George and Millie *Frank Gallo*	145
14	Dedication of the Hawk Watch Plaque at Lighthouse Point Park *Arne Rosengren*	157

15	The Birdwatchers *Owen Elphick*	163

PART VI
Notes from the Field   165

16	Bounty Hunter *Frank Mantlik*	167
17	Through the Lens *Lesley Roy*	175
18	Eagle Rock *Deborah Johnson*	191
19	Brainy Birds *Gail Martino*	199

Appendices   209

THE NEW HAVEN BIRD CLUB AND ITS PROGRAMS	211
TWENTY-FIVE FUN FACTS ABOUT THE NEW HAVEN BIRD CLUB	217
ABOUT THE AUTHORS	221
ACKNOWLEDGMENTS	227

Preface

An incredible number of people enjoy birds in the United States and in my state of Connecticut. The most recent survey results by the US Fish and Wildlife Service reveal that around 45 million people in the United States currently consider themselves birdwatchers. About one million of these people live in Connecticut, making the Nutmeg State sixth in the nation in birding popularity. A subset of people fascinated by these winged wonders have shifted from merely watching birds pass by to searching for birds—and so are typically termed "birders." I am one of these folks. In spring, summer, fall, and winter, I can be seen with binoculars or scope in hand, scanning the trees and horizon for winged treasure.

As one of the compilers of *Chickadee Tales*, I owe you a brief description of some of my own personal history. My professional life is in a scientific field. I relocated to New Haven over a decade ago to work for my current employer. At that time, my assumption about living in Connecticut was: "This will be fun for eighteen months." All these years later, I am still living in New Haven. One reason is that residing here has afforded me the opportunity to broaden my birding skills. As a member of the New Haven Bird Club (NHBC) and as a board member, serving as Indoor Programs Chair, I have met many active and skilled birders. My thanks go out to them for sharing their experiences, knowledge, and personal stories with me—all of which fueled my desire to see this book reach the finish line.

In my role as Indoor Programs Chair of the NHBC, I organize a monthly speaker series as well as other educational activities, such as tours of the ornithological collection at Yale University's Peabody Museum of Natural History. One of my greatest pleasures in this role is seeking out and listening to birding-related stories of other NHBC members.

In my own case, I could say that viewing a particular bird—and my father's gentle guidance as a casual birdwatcher—changed my interest from an early curiosity to a lifelong passion for birding. It began for me as a child, in the greater Boston area, one spring day when a nearly robin-size bird with a black cowl and contrasting orange breast alighted atop the Weeping Willow tree in our backyard. I listened intently to its liquid-sounding warble. When I asked my father about this jewel of a creature, he informed me that it was a Baltimore Oriole and it lived in our backyard during the spring and summer months.

Fascinated by this bird in our yard, I continued to watch its activities throughout the spring and summer. I heard the male oriole sing as he established his territory and watched his mate weave the species' distinctive, pendulous nest. The Baltimore Oriole's life was far more intricate and complex than I had imagined.

Several decades later, I have retained that sense of wonder when observing birds and hearing stories about them. One person who shares my appreciation for these stories is Ricci Cummings, my companion compiler for *Chickadee Tales*. A retired lawyer, Ricci tells her own story regarding her interest in birding in the essay "Why We Bird." Ricci and I are excited to offer the local and national birding community this anthology of essays (the first anthology published by the NHBC). In compiling this volume, our aim was to provide a mix of informative and entertaining stories, presented by NHBC members and supporters.

Regardless of whether you are a novice or an experienced birder, we hope you will enjoy reading this anthology as much as we enjoyed assembling it for you. *Chickadee Tales* is for anyone enthralled with birds, nature, conservation, and/or connecting with one's "flock" though bird-oriented clubs.

—Gail Martino

Chickadee Tales:
An Introduction

The city of New Haven, Connecticut, may not leap to mind as a prime birdwatching environment. It is, however, a key New England birding destination. With beaches (to the south), cliffs (to the east and west), and large areas of deciduous woods and marshlands, the New Haven vicinity provides year-round enjoyment of birds in a variety of habitats. If you are a bird enthusiast living in a different area of the United States, you may be surprised to discover that the New Haven area is blessed with some of the most productive birding areas not only in Connecticut but along the entire East Coast.

In the springtime, New Haven birders may observe colorful neotropical warblers on their northward journeys at East Rock or West Rock parks. In autumn, raptors migrating southward can be viewed at Hawk Watch in New Haven's Lighthouse Point Park. Over the course of a year, a skilled birder might view at least two hundred species of birds in New Haven alone—and over three hundred in Connecticut—so it's a terrific place to begin or expand your experience as a birder.

As if this were not sufficient argument for planning a birdwatching trip to this south-central part of Connecticut, New Haven boasts higher-education institutions, including Yale University, Quinnipiac University, and the University of

New Haven, where you can broaden your understanding of ornithology.

Once in New Haven, consider connecting with members of the New Haven Bird Club (NHBC). The NHBC has extensive experience in welcoming bird enthusiasts and creating and maintaining a bird-loving flock. One of the first birding clubs in the United States, established 113 years ago, the NHBC currently has around five hundred members. Its diverse activities include birding walks and indoor and virtual presentations by leading birders, as well as fieldwork opportunities, competitive and noncompetitive birding events, and community conservation and educational activities. Indeed, NHBC members are people like you who are enthralled by birds and want to connect with others who share this passion.

Chickadee Tales: **What's Inside**

This book's title, *Chickadee Tales*, stems from the NHBC's logo, a Black-capped Chickadee encircled by the club name. The bird is one of New Haven's year-round residents. With its friendly and inquisitive nature, the chickadee is often observed among flocks in woodlands, much like our members!

This collection of essays, contributed by NHBC members and supporters, represents some of the diverse interests and experiences of the club's constituents. You will find NHBC historical accounts as well as reports of first-person encounters with birds, conservation efforts, and how NHBC members have mentored younger generations of birders. Some of the essays are deeply personal, others more science-oriented. By reading the nineteen accounts and bonus appendix included in this anthology, we hope you will gain a better understanding of the varied motivations, activities, discoveries, personal connections, and contributions of NHBC members and supporters. In keeping with the present-day nomenclature of the NHBC, people like us—who

An Introduction

are avid about birds—are termed "birders" herein, rather than merely "birdwatchers."

Both professional writers and first-time authors contributed essays to the six different parts within this anthology. The vocations of some of the contributing writers are related to birding. Others have professional lives outside of the realm of science and nature but are fascinated by birds and birding. Each of the authors was granted the complete authority to develop and express his or her own content and style. In this way, the co-compilers (Gail Martino and Ricci Cummings) were able to capture a sense of the range of thoughts, ideas, and voices that compose the NHBC as an organization.

The essays included in this anthology have been grouped thematically. Part I: Foundations contains the essay "A Century and More," in which former NHBC President Craig Repasz conducts a tour of the club's history, from a December morning in 1906 to the present day. Written with input from the NHBC historian, John Triana, and other seasoned members, this essay describes the club's origins and offers both historical and present-day perspectives.

Part II: Birder Beginnings conveys particular experiences that launched various members' enthusiasm for birding. The essay "Why We Bird," written by Gail Martino, starts with the author's answer to the question, "What got you interested in birding?" Following this is a synthesis of interviews with new and experienced birders, who reflect on that same question. In "Interview with Arne Rosengren," NHBC member DeWitt Allen transcribes an interview between Florence McBride and her friend Arne Rosengren, one of the club's oldest members, about his lifelong interest in birding. If you are a birder and/or involved in some other bird-focused club or organization, these foundational stories may resonate with your own experiences as they give you a broader sense of others' formative experiences.

The four essays in Part III: Special Programs afford the reader insight into the experience of birding by describing the sights, sounds, thoughts, and actions stemming from some of the NHBC's signature programming and activities. "Ed's Count," by Frank Gallo, is a fictionalized narrative of one person's experience of the annual Christmas Bird Count, a national event in which the club has participated since its founding. Steve Mayo's "Hawk Watch" outlines the fifty-year history and the soaring delights of the club's annual tracking of the autumn raptor migration through New Haven's Lighthouse Point Park. Chris Loscalzo, who debuted the Mega Bowl of Birding for the NHBC, gives a birder's-eye view of this annual competitive event. Ricci Cummings offers a perspective on the club's many birding walks through the lens of the First Wednesday Walks organized by the dedicated birder Tina Green.

Part IV: Education and Conservation includes case studies of meaningful contributions to birds and birding by NHBC members. In "*Take Flight!* Bird Observation with Children in a School Science Program and Beyond," Florence McBride, long-time educator and leader of children's bird walks, outlines an award-winning school program designed to interest youngsters in birds and nature. "Restoring Local Populations of American Kestrels, One Box at a Time" is the story of one man's decision to address the challenge of declining populations of kestrels in Connecticut. Tom Sayers, founder of the Northeast Connecticut Kestrel Project, describes how he went from knowing very little about these falcons to making a significant contribution toward their recovery in the state. Pat Leahy, in "A Condo Development That Bluebirds Love," details his successes in providing nest boxes for and tracking the reproductive successes of the Eastern Bluebird in Connecticut. All these essays demonstrate that one person can make a difference.

An Introduction

Part V: Tributes highlights contributions of NHBC members who have particularly impacted our birding community. Written by former colleagues, students, and friends, "There's Always Something to See: A Tribute to Noble Proctor" pays homage to the late, beloved Professor Emeritus of Biology at Southern Connecticut State University. "The Gifts of Richard English," reported by Martha Lee Asarisi and Michael Horn, describes the lifetime and posthumous gifts of Dick English, who enabled so many people to get closer to nature and birds. As told by Frank Gallo, one of their close friends, "George and Millie" portrays the impact of George and Millie Letis on the birding community. "Dedication of the Hawk Watch Plaque at Lighthouse Point Park" is a transcription of a talk given by Arne Rosengren in 1994. This section concludes with a poem Owen Elphick wrote to honor his father, Chris Elphick, a birder and University of Connecticut Professor of Ecology and Evolutionary Biology.

In the book's last part, "Notes from the Field," several NHBC members recount personal experiences with birds and birding. In "Bounty Hunter," Frank Mantlik depicts the search for rare birds as a member of the Avian Records Committee of Connecticut. Lesley Roy's "Through the Lens" recalls a specific encounter that led to the pursuit of bird photography. Deborah Johnson's "Eagle Rock" explains how a birding excursion jump-started a relationship that led to marriage. Finally, Gail Martino's "Brainy Birds" notes how even the simplest observation at a backyard feeder can spark curiosity about bird behavior and cognition.

Even if you are new to birdwatching, we hope that the contributions and contributors in this anthology speak to you on a personal level and that when you finish reading this book, you emerge with a new or enhanced appreciation of nature, birds, birding programs, and the value of a bird club and/or your fellow flock mates. Most of all, we hope you will consider (or continue) birding in the New Haven area—a fabulous place to view and interact with both birds and bird-lovers!

PART I
Foundations

A Century and More

Craig Repasz

The Bookcase

"You need to see this," John Triana exclaimed.

The past President and current Historian of the New Haven Bird Club (NHBC) and I had spent a few fascinating hours discussing the club and the parade of individuals who had contributed to its history. John was beaming with pride as he pointed to a bookcase loaded with books about birds and birding. A mix of volumes through the ages, it held field guides by David Allen Sibley and old volumes by forgotten names like Herbert K. Job. I knew it was going to be a very late night.

John Triana was President of the NHBC from 2003 until the club's centennial, in 2007. During this time, he had compiled a large number of published books, scrapbooks, photographs, minutes, and clippings. Many of these materials were drawn from the club's archives; others he compiled by contacting descendants of past members. He gave three presentations of his material at the club's indoor meetings, in 2006, 2007, and a reprise in 2018.

John explained that he was maintaining a collection of books written or illustrated by people who had been members of the NHBC. Looking at the authors in this impressive collection, it is evident that a history of birding, ornithology, and bird conservation in Connecticut and the United States could be written through the biographies of the club's members. Some of these authors and artists walked with US presidents, traveled to distant lands, and captured the essence of a bird with pigment on paper. Although these volumes on John's shelves, tucked under the sloping ceiling on the second floor of his quaint Cape-style house in Bethany, could be measured by yards, they still came up short in revealing the full story of the NHBC. This essay will try to tell that story, which starts in the shadows of Yale University's Gothic-style buildings in New Haven, the largest city in Connecticut at that time.

This small local organization, now in its 113th year, first appeared as a gathering of teenagers, a product of a nationwide nature-loving moment that saw many similar associations spring into existence. The club utilized the power of women, before they had the right to vote or hold office, to be leaders and a driving force. Many leading artists, conservationists, and ornithologists appeared on the club's membership lists, and many live on in John's bookcase. The story is one of fellowship, traditions, and passions that have defined the club and carried it for a century and more.

Beginnings

On a cold Christmas Day in 1906, five young men stepped out into the winter landscape around New Haven. Flurries had coated the ground with a few inches of snow. The men were dressed in woolens, canvas, and leather to face the twenty-degree temperatures. They could have been figures captured half a century earlier by George Henry Durrie in one of his bucolic Connecticut winter scenes. However, they were not in horse-drawn sleighs nor on their way to church. They were stepping out to count birds and, breaking from a long-standing tradition, were doing it without shotguns.

Aretas A. Saunders lived five hundred feet from Edgewood Park, in a part of New Haven where the streets had branched out from the city's original nine squares and now awaited the building of houses. He had been surveying the park for the Christmas Bird Count since 1901, the second year of the annual avian census. Starting in 1905, Saunders would be joined by friends from the developing Edgewood neighborhood—Clifford and Dwight Pangburn and Bernard Leete—while another friend, Albert Honywill, lived a little farther away, near the Grove Street Cemetery. Albert could either walk to join his naturalist companions or jump on the Whalley Avenue trolley.

The trolley could bring these young men out of the largest city in Connecticut to the edge of the wilderness, to the beaches and marshes of Long Island Sound, or to the hills, fields, and forests that surround the city. On this day, they were able to cover Lake Saltonstall, Lighthouse Point, Momauguin (East Haven), Edgewood Park, Mitchell's Hill, West Rock, Lake Wintergreen, Prospect Hill (rising above Yale's Sheffield Scientific School, now Sheffield-Sterling-Strathcona Hall, on Prospect and Grove Streets), and the swath from Beaver Ponds to Pine Rock. All totaled, these five young birders covered six routes over three days, splitting up individually or into teams. While his friends were off surveying other parts of the city, Clifford Pangburn ambled west through Edgewood Park toward Marvelwood and Mitchell's Hill (now the Yale Golf Course and the Brooklawn Circle area). The area, *Connecticut Magazine* claimed in a 1904 article, had remained wild even after the founding of New Haven Colony in 1637.

At Clifford Pangburn's back, the city of New Haven, nestled under the bare boughs of stately elms, its factories silent, was waking to Christmas celebrations. The coal-fired furnaces were just getting stoked in the houses. The cliffs of West Rock, once called Providence Hill, hung frozen to his north and northeast. Back in 1661, Edward Whalley, a British military officer serving under Oliver Cromwell, and his son-in-law, William Goffe, hid on West Rock ridge in a pile of glacial erratics now known as Judges Cave. Two of the judges that had sentenced King Charles to be beheaded, they were avoiding arrest by deputies of the new king, Charles II, and certain execution in England. From this vantage point, the men could look down on the tiny hamlet of Westville and the very expanse of wilderness where Clifford Pangburn, 245 years later, would count a winter flock of forest birds.

Clifford Pangburn and his friends, although aware of the history hanging over them, were not aware of the forces already in play that would shape their surroundings, or a conservation movement in its infancy, or their role in the story now being told, more than a century later.

Four months after this Christmas Bird Count, the NHBC would transform from an informal fellowship of boyhood friends to a strong organization that would influence the birding and conservation worlds for over a century. These young men had already completed a constitution and bylaws, signed by four of them with silly bird cartoons accompanying their signatures. From these youthful beginnings, the NHBC would soon expand to include students from Yale and the local high schools, pastors, writers, artists, conservationists, lawyers, educators, doctors, and professional ornithologists.

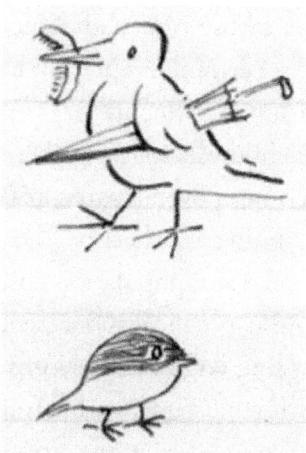

Signature illustrations from the constitution and bylaws drafted before the New Haven Bird Club's official founding. NHBC collection.

Just a few years after the Pangburn brothers' Christmas Day excursion, the wild areas beyond Edgewood Park would begin to succumb to the suburban sprawl that would consume the eastern seaboard of the United States. This growth, called "progress" and "development," was soon invigorated by the manufacture of automobiles and the endless supply of cheap fuel. A rising middle class and the creation of leisure time would give individuals and families the wherewithal to take up the hobby of birding. People found that they had disposable income to purchase the books and optics, and transportation so they could survey birding hot spots throughout the county. The very factors that allowed the NHBC and like organizations to prosper were also the factors that threatened our beloved birds and their habitats.

The National Scene

There was nothing unique about a local bird club starting up in a city during the first decade of the twentieth century. Such organizations were springing up across the country like mushrooms popping through the forest floor after an autumn rain. A local bird club was an idea whose time had come. Although not original, the NHBC was certainly representative of a grassroots nature and conservation movement.

The club was the child of a marriage between the nature-loving moment and a rising movement to preserve the nature that Americans were beginning to cherish. The nineteenth century saw the decimation of many populations of plants and animals on the American landscape. Professional scientists and amateur collectors were providing large numbers of eggs, bird skins, and preserved specimens for museums and private collections. Some private egg collections numbered in the tens of thousands. Market hunters, with no bag limits to restrict their take, were bringing down whole flocks to sell in food markets. The Passenger Pigeon, once traveling in flocks of hundreds of thousands that clouded the skies for days, plummeted to extinction. Ornithologist Alexander Wetmore reported the last wild sighting in Kansas in April 1905. The last Passenger Pigeon, named Martha, died in the Cincinnati Zoo of old age and loneliness on September 1, 1914.

There was slaughter for slaughter's sake: hunters shot entire herds of American Bison, only to leave their carcasses on the prairie to rot. Groups would participate in the "Side Hunt," the precursor to the Christmas Bird Count, in which competitors would shoot birds and measure their pile of carnage against their rival team's pile. Women were wearing bird feathers and entire stuffed birds in their hats, fueling a demand that was met by further carnage. With reports of the damage becoming widespread, the once-held belief that nature's bounty was infinite and at the developed world's disposal was challenged. Conservation efforts were well underway by the end of the 1800s.

The beginnings of the conservation movement have a New Haven connection in Yale graduate (class of 1870) George Bird

Grinnell, who would launch a nationwide effort that would fail. Among the ruins, a new movement would emerge, based in cities and states; the NHBC was part of this second wave. The economic panic in 1873 had pushed Grinnell from his family's New York investment bank to a position as an assistant to the era's preeminent paleontologist, Othniel C. Marsh, at Yale University's Peabody Museum of Natural History. Grinnell would become an avid collector and student of natural history of the American West. He studied the effect of unlimited hunting on mammal populations in Yellowstone Park and joined one of General George Custer's expeditions into the Black Hills of the Dakotas, collecting specimens to be sent back to the Peabody as well as other eastern museums. Custer invited Grinnell to accompany his cavalry unit on a second expedition in 1876. Grinnell declined, choosing to return to Yale to continue his studies, and so avoided becoming another casualty of the Lakota, Cheyenne, and Arapaho at the Battle of Little Big Horn.

By 1880, Grinnell, now with a PhD, was in the publishing business; he and his father obtained the rights to the conservation periodical *Forest and Stream* (which later merged with *Field and Stream*), and he became senior editor and publisher. Grinnell used the magazine as a conservationist mouthpiece and soon found another forum, the American Ornithological Union (AOU), formed in 1883 by shotgun-toting ornithologists. Grinnell and the AOU could agree on a common enemy, the milliners and their insatiable demand for fashionable feathers. However, the AOU was not unified in its voice against the feather trade. Some of its members felt that any action against the trade might be extended, posing a threat to their favorite taxidermist and their own collecting. After a good start, the AOU harrumphed to a standstill by 1890.

Grinnell was not satisfied. To expand his defense of birds, in 1886 he proposed the formation of the Audubon Society, for the "protection of wild birds and their eggs." He chose the name because of his personal connection to John James Audubon's widow, Lucy, whom he had met as a boy at Audubon Park in New York. With support from Henry Ward Beecher, John G. Whittier, and John Burroughs, all prominent dignitaries from the nature movement, the organization gained momentum.

John Burroughs, especially, would impact the newly organizing birders in New Haven. A native of the Catskills region of upstate New York, Burroughs claimed Walt Whitman as a dear friend. In his later years, he was acquainted with and went on excursions with many personalities of the time, including Theodore Roosevelt, John Muir, Henry Ford, Harvey Firestone, and Thomas Edison. In 1899, Burroughs participated in E. H. Harriman's scientific expedition to Alaska, where he met Leon J. Cole—who later became a member of the NHBC and a pioneer in bird banding. Burroughs wrote extensively about nature, religion, philosophy, and literature, offering commentary on the works of Walt Whitman, Ralph Waldo Emerson, and Henry David Thoreau, among others.

In the Catskills, Burroughs spent much of his time observing birds, insects, fish, and trees. He advocated escaping the musty museums and stuffy libraries to get out into nature to experience it firsthand, becoming the voice in the wilderness calling others to come out. Burroughs was not opposed to the taking of birds and their nests and eggs as part of the process of learning about them—as long as this was done in concert with the observation of live birds and their behaviors and listening to their songs. After mastering the birds in this manner, he writes, in his essay collection *Wake-Robin* (1871), "the true ornithologist leaves his gun at home."

In the mid-1880s the conservation movement seemed strong. Although the AOU was dominated by the specimen collectors in its ranks, the organization had managed to draft model laws for bird protection that were picked up by several states and municipalities. These laws were in effect only at local levels, when it was clear that national laws were needed.

By the end of the century, Grinnell and his Audubon Society, like the AOU, had lost their momentum and stagnated. The Audubon Society had never collected dues or subscriptions, asking only for donations. With a lack of funding and an overworked staff, the society stopped publication of *Audubon* magazine. Furthermore, indifference and apathy would overcome the AOU's ranks. With the loss of their champions, North American bird populations would continue to suffer from specimen and egg collecting and the ravages of the feather trade. The men had had their chance. Now was the time for the women to take up the baton.

It took a Boston Brahmin, reading an article about the slaughter of birds to feed the feather trade while sitting over her afternoon coffee on a cold winter day in 1896, to spark a conservation movement that rages on to the current day. Harriet Lawrence Hemenway, fueled with outrage, was able to combine her organizational talents with her passion to protect birds and create the Massachusetts Audubon Society (now known as Mass Audubon) in a short period of time. The movement did not stop at the Massachusetts state line; soon there were state and local conservation organizations throughout the United States.

This second attempt to galvanize a conservation movement succeeded where previous attempts had failed. The reasons for this success were many: There was now a focus on local and state levels reaching up, not a national level reaching down. People were ready; there was a vast popular interest in communing with nature and getting outdoors, in large part owing to a wealth of nature writers such as John Burroughs. This new movement stressed educational programs. Soon the state and local organizations were consolidated into the National Association of Audubon Societies, which worked with the resurrected AOU on conservation efforts.

Most of the new Audubon state and local organizations were being driven by women, but they needed men to front them in an age when women could not vote. Mass Audubon elected William Brewster, who had been a founding member of the AOU, as its first president. Other men who became prominent conservation voices in this period were Clinton Hart Merriam, William Dutcher, and Frank M. Chapman. Merriam, a biologist who had studied at Yale's Sheffield Scientific School, founded the National Geographic Society and what is now the US Fish and Wildlife Service. Dutcher, the AOU's treasurer, held a passion for bird conservation and pushed forward a new effort. His initiative lacked wide support within the AOU, and the committee handling it was overwhelmed with inquiries and correspondence. Frank Chapman, an ornithologist, felt that a lay magazine about birds and bird conservation could take the burden of correspondence off Dutcher's overworked committee. In 1897, Chapman launched the national periodical *Bird-Lore*, the "Official Organ of the Audubon Societies." Chapman turned to Mabel Osgood Wright, of Fairfield, Connecticut, to be his editor. Wright founded the Connecticut Audubon Society in 1898 and quickly became a prominent voice in the conservation movement.

It was in *Bird-Lore*, in 1900, that Chapman first proposed a simple idea: rather than going out to shoot birds on Christmas Day, people should go out and count the birds in their area instead. Connecticut birders participated in the first Christmas Bird Count in 1900. Aretas Saunders would conduct count surveys in Edgewood Park in most years leading up to the founding, in 1907, of the NHBC. In 1902, Saunders tallied nine species in Edgewood Park, noting the presence of Eastern Bluebirds. In 1903, he recorded a Northern Shrike. In 1905, he was accompanied by the Pangburn brothers and Albert Honywill. In 1907, Mrs. C. A. Dykeman joined the core group of young men. In 1908, they experienced an exciting flight year of boreal finches and crossbills.

These young men and women were riding the new crest of nature observation. Nature education was at the forefront of the newly emerging Audubon groups. Books on how to observe and identify birds in the field had started to appear in the late 1880s. In 1890, Florence A. Merriam (Clinton Hart Merriam's sister) published *Birds through an Opera-Glass*, which coaches non-ornithologists, using low-magnification glasses for field observation, in what to look for to identify birds.

Many people were birding with opera glasses or old army field glasses, both of which were two telescopes in tandem with a central focus. These crude optics were the norm until 1893, when Ernst Abbe, a partner in the Zeiss company, invented Porro prism binoculars. These new binoculars provided magnification ranging from seven to ten times the original, while opera glasses offered a magnification factor of around three. Such optics, however, were expensive and accessible to only a select few and, hence, would not impact birding until decades later, when they became more affordable.

In 1882, years before Florence Merriam published her book, she became a student at Smith College in Northampton, Massachusetts, where she organized nature walks led by John Burroughs. These walks helped create a long-standing culture of nature studies at Smith College. Ida Barney, a 1908 Smith graduate, became the president of the NHBC from 1956 to 1958.

In 1895, Mabel Osgood Wright, a contemporary of Florence Merriam, published what many considered the first modern field guide, *Birdcraft: A Field Book of Two Hundred Song, Game, and Water Birds.* Wright illustrated the book with reproductions of works by John James Audubon, Louis Agassiz Fuertes, and other artists. Her book focused on common birds found in urban parks and neighborhoods, which she described and presented to a popular audience. Frank Chapman described it as "one of the first and most successful bird manuals."

This inexpensive volume filled a void for American birders and was praised by the likes of John Burroughs. It was reprinted nine times and served as the most popular bird guide until Roger Tory Peterson's *A Field Guide to the Birds of Eastern and Central North America* was published in 1934.

A Bird Lovers' Club in New Haven

The first decade of the twentieth century was the perfect moment for the formation of the NHBC. In 1907, the *New Haven Register* claimed that the interest in bird study was at "fever heat." A sustaining wave of popular nature writing was getting people outside; there were crashing conservation crises that needed passionate supporters; and a flurry of organizations was forming to educate people about nature and champion conservation causes. The young men who were out doing the 1906 Christmas Bird Count would soon join others in New Haven: educators, ornithologists (both professional and amateur), and women advocates. They had a choice: they could join the nine-year-old Connecticut Audubon Society, founded by Mabel Osgood Wright, or they could go it alone and start their own organization. In 1907, the NHBC came into existence, choosing to go it alone.

The *New Haven Register* ran an invitation on March 26, 1907, "cordially extended to teachers, students and in short all interested in the study of our native birds," to come to an organizational meeting in the cold hour of 8:00 p.m. on March 27 to start a club. The widespread interest in birds throughout the city, including the incorporation of nature study into the public-school curriculum, had prompted the announcement. The primary targets for members were educators; the new club hoped to enroll as many teachers as possible, so that its members could help them with their school programs. Already the club was forward-thinking, looking beyond the city of New Haven to include surrounding towns in the newly forming organization.

That first meeting, held at the New Haven YMCA, was attended by fifty people. An ad hoc committee was formed, the participants of which would become prominent club members: Edgar Stiles, Philip Buttrick, Elizabeth Whittlesey, Alfred Kedzie, and Aretas Saunders, the stalwart birder of the city's early Christmas Bird Counts. The committee was entrusted to draw up a plan for a permanent club, and to choose an appropriate name, before the attendees were to formally meet again in a week.

The group reconvened on April 3, this time drawing sixty people to the YMCA. The meeting was called to order at 8:25 p.m. Each article of the constitution and bylaws drafted by the committee was read and accepted. The constitution's first line read, "The name of this club shall be the New Haven Bird Club." The documents defined the scope of the club to be "about the same as that of the numerous Audubon Societies throughout the country. Bird study, however, is the main object of the organization, but bird protection will also be sought." The minutes record that a nominating committee was formed, and officers were elected: Edgar Stiles, president; Alfred S. Kedzie, vice president; Aretas Saunders, recording secretary; Estelle (Mrs. Walter C.) Greist, corresponding secretary. The club planned to meet every month at the YMCA. The attendees agreed to dues of fifty cents for the year, payable at the annual meeting. A *New Haven Register* article states that the club's new members came from "all walks of life, teachers, students, artisans, lawyers, businesspeople, and other professional men and included many women." The women "evince[d] an enthusiasm that seem[ed] to exceed that shown by the male members" and vowed to campaign against the feather trade driven by the demand for fashionable hats.

Definite plans for the club's activities were not drawn up that evening; however, the members were determined to have lectures on avian topics and to correspond with like organizations in the country. A large component of the club's meetings would be the reading of ornithological papers and correspondence. Attendees expressed hope that in the future there would be a club room to hold documents and photographs and give the members a space in which to look over these materials. During winter months, when birding tends to be slow, the club planned to view pictures taken during the previous summer on lantern slides and to host a discussion on winter birds.

Immediately the club began organizing outdoor trips, the *New Haven Register* reports, "to the haunts of the feathered tribe," with an emphasis on spring migration and the timing of the birds' arrival in New Haven, where they would be observed by "the disciples of Audubon and Burroughs." These observations would be exchanged with other birding organizations in the country, so that a large picture of migration could be filled in.

The meeting adjourned at the late hour of 9:50 p.m., and the members stepped out into the cold April night. The air held the promise of spring and the return of the city's migrating and breeding birds. This spring's arriving warblers would be greeted by the new club.

The founding date of April 3, 1907, was serendipitous; it was also the seventieth birthday of writer-naturalist John Burroughs. The club voted to extend honorary memberships to Burroughs, whom Kedzie had visited the year before in New York, and to Donald Grant Mitchell, another famous writer of the day, best known by his nom de plume, Ik Marvel. Mitchell had donated a large part of his estate to the city of New Haven in 1889, to become part of Edgewood Park, and had sold another portion to New Haven businessman John Greist, who named his estate Marvelwood; it was on this massive property that Saunders had been conducting his Christmas Bird Count since 1902.

The club's formation was lauded by the community. Although the founding ideologies were bird study and conservation, an observer noted how wonderful it was to have an organization that could help ascertain the value of birds to the economy: How many insects would they eat? How many weed seeds would be consumed? What savings to agriculture? This economic approach was giving way to the more fundamental ecological thinking that nature must be preserved for its own sake. The early NHBC was not interested in how many weed seeds the local finches would consume. Its members were fascinated that the finches were in the city, and they would work to keep these and other birds around.

Two weeks later, on April 17, the new organization—referred to in the press as "the Bird Lovers Club"—met again. The charter members were a serious bunch who understood the need to have an organization with an established constitution and bylaws. However, they did not spend much time on these details, choosing instead to tackle the mission of being a bird club. They did note, with eighty members now in attendance, that they needed to find a bigger room for their meetings. Mrs. Greist read a letter from John Burroughs accepting his honorary membership.

John Shepard gave a talk, "How to Learn the Birds." A question-and-answer period that followed moved to a discussion of how important it was to learn and have a fundamental knowledge of bird songs. The spring migration was upon them, one attendee pointed out, but the reasons and mechanisms for migration were unknown. In response, the club formed a task force to list all the birds that were migrating through the area. The immediate need to record their observations was underscored by the report that one member, Estelle Greist most likely, had listed eight birds singing within thirty seconds near Mitchell's Hill. The spring migration was on.

Early Initiatives

One of the early objectives of the NHBC was to harness the youthful energy of its student members, including Saunders, Honywill, the Pangburn brothers, and Leete. They would patrol Edgewood Park, policing nests and eggs of the resident birds to ensure they would not be destroyed or taken by vandals, hunters, or collectors. Schoolchildren would be permitted to collect only deserted nests. These patrol efforts were in response to the sentiment, expressed in the *New Haven Register*, that boys could be a nuisance and threat to local birds. With a little effort, it was hoped, these local boys could be converted into friends and protectors of the birds. The *Register* reported that the real threats to the local bird population, even after decades of assaults by collectors and market hunters, came from colonies of feral cats in the city parks and from new Italian arrivals, who were accused of bringing with them a custom of eating small birds.

While the club's youths were taking to the parks, the women were beginning to add their voice to the national movement to stop the feather trade. They went door-to-door to the shops in New Haven that were selling hats adorned with aigrettes, or sprays of feathers. They distributed pamphlets that educated consumers and ran articles in the local newspapers. They found that people were unaware that the plumes were taken from egrets and other birds shot while tending to fledglings in the nest. These hunters would take the best feathers and leave the dead bird behind. To make matters worse, the entire nest of young would soon starve or succumb to predation.

As these initial efforts to promote education and conservation were ongoing, club members took to the streets and parks to observe visiting and resident birds. They compiled their observations, carried on correspondence with other bird clubs, started bird-banding projects, and created an early atlas of the New Haven region to indicate where the birds were to be found. In 1908, the original young Christmas Bird Count participants—Saunders, the Pangburn brothers, and Honywill—joined up with Freeman Burr, Philip Buttrick, and Dr. Louis Bishop to create a checklist of the New Haven birds and a description of the club's favorite birding hot spots.

The club's initial twenty-five years were very busy and expansive. Membership grew to 125 in the first few years, and the club squeezed out of the YMCA building and into the Normal School, a teachers' training facility on Howe Street. By the end of its first quarter century, the club had begun to hold its meetings at the Yale Peabody Museum. After a couple of years of calling the Peabody Museum auditorium home, the club bounced from the Peabody's laboratories on Hillhouse Street to the Yale Forest School's (Yale School of the Environment) Sage Hall, then to various churches, Yale's Osborn Memorial Laboratories, across from Ingalls Hockey Rink, to Quinnipiac College (now Quinnipiac University) in neighboring Hamden, the Connecticut Agricultural Experiment Station, and finally the Whitney Center on Leeder Hill Drive in Hamden.

Other bird clubs came into being around the state at about the same time as the NHBC. These clubs formed the Connecticut Federation of Nature and Bird Clubs in 1917, based in "the Fairfield Bird Sanctuary" (the present Birdcraft Sanctuary). In 1921, the NHBC hosted the federation's annual meeting, arranging field trips to West Haven Sand Spit, West Rock Park, Edgewood Park, and Marvelwood. The business meeting covered field reports on birdsong, ferns, and morning walks.

Early Influential Members

NHBC members, some of them Yale faculty or students, have made powerful contributions in many fields. Charter member and Yale grad Ernest Coe was instrumental in the establishment of Everglades National Park. Club President Ida Barney made her mark in astronomy, researching the stars for three decades at the Yale University Observatory. Other members made an impact on the state and local level. Edgar Stiles had a lasting impact on education in Connecticut, while Alfred Kedzie wrote about birds in the local papers. Among the many remarkable men and women, several early members stand out for their contributions to birding and their representation of historical themes in the history of birding and ornithology.

Dr. Leon J. Cole joined the young club in 1908. He had received his PhD from Harvard in 1906 and was already established in the fields of animal breeding and pathology. He would later go on to become an outstanding geneticist. Dr. Cole had an immense love for and fascination with birds. In 1901 he had written an article proposing the idea of using leg bands on birds to give observers a means to track migration. This article, and his presentation to the AOU the same year, is seen as the first proposal for a comprehensive bird-banding program.

Immediately upon joining the NHBC, Cole formed a bird-banding committee with Dr. Louis Bishop and Clifford

Pangburn. The committee developed procedures for banding, including how to collect and retain data. These efforts soon expanded beyond the NHBC, becoming national in scope after a report to the AOU in 1909 resulted in the formation of the American Bird Banding Association in New York. Dr. Cole served as the association's president, wrote a number of papers on bird banding, and is regarded as the father of bird banding in the United States. Cole left New Haven in 1910 to take a faculty position at the University of Wisconsin and set up the Department of Experimental Breeding for plant and animal improvement. His new endeavors kept him too busy to pursue his interests in bird banding, but his efforts were not forgotten. In 1928, Dr. Frederick Lincoln, in a history of bird-banding in the United States, wrote, "To Dr. Leon J. Cole must go the credit, however, for bringing the advantages of the method [i.e., bird banding] to attention of American ornithologists."

A Congregationalist minister, NHBC member Herbert K. Job considered birds to be a creation of God and believed every person had a holy obligation to protect them. Although zealous in his beliefs, Job had the temperament of an artist. Born in Boston, he received a BA from Harvard and served as minister to a congregation in Massachusetts before he moved to Connecticut. His passion for birds drew him to their study and to photography and film. He took a faculty position at the Connecticut Agricultural College (later to become the University of Connecticut) from 1908 to 1914, and during this time also served as the state ornithologist. Job focused on economic ornithology from 1914 to 1924, a field of study that viewed birds as either a value or a risk to agricultural economies. This approach to commodify birds stands in contrast to his spiritual belief that birds are a gift from God and to his artistic efforts to capture birds on film.

A voice for bird protection, Job called for a move away from hunting, offering shooting birds photographically as an equal challenge for what he called "camera-hunters." His early work with bird photography helped to bridge the collectors of the nineteenth century and the artists and photographers of the future. In the early 1900s, while the carnage of the feather trade was strong, he took trips throughout the eastern United States, with an emphasis on Florida, to photograph birds. Fellow NHBC member Dr. Louis Bishop was a frequent companion. Job's book *Wild Wings: Adventures of a Camera-Hunter among the Larger Wild Birds of North America on Sea and Land* (1905), written from his home in Kent, Connecticut, describes his awe for birds. The kittiwakes and gannets "and their lone surroundings," he wrote, "are the unsullied handiwork of God. No trace is here of man's vandalism; the wildness of the scene might well have been matched at the Creation's dawn." The book's preface includes a note from Job's friend and Harvard schoolmate Theodore Roosevelt, applauding the importance of the move from gun to camera.

Job's trips to Florida and concern for the protection of birds led him to push for the preservation of the Florida Keys. Responding to Job's and others' influence, Roosevelt created the Key West National Wildlife Refuge in 1908, one of the first refuges in the country. In June 1915, Job accompanied Roosevelt on a visit to the beaches of Louisiana, taking photos and making movies. Job published seven books on birds and made three movies.

Another prominent NHBC member, Dr. Louis Bishop, was born in his maternal grandparents' home in Guilford, Connecticut, in 1865, two months after the Civil War ended. His family descended from Connecticut's original English colonists. Bishop grew up on Church Street in New Haven and attended the Hopkins Grammar School. As a boy he would shoot birds with a slingshot and make skins for his collection. When he was twelve years old, he received formal training in skin making and specimen collection from a Yale student, who taught him for a fee. Bishop had to supply the birds. His collection would grow so large, he built a small museum on his property to house it. In 1939, he donated fifty-three thousand specimens to the Field Museum of Natural History in Chicago; other skins and eggs are housed at the Yale Peabody Museum.

Bishop attended Yale, receiving a BA in 1886 and a medical degree in 1888. While a twenty-year-old student, he published his first ornithological journal article with the AOU. He established his residence and medical practice on Orange Street in New Haven. Although Bishop was a successful doctor, ornithology was his true passion. In 1908 he retired from his medical practice to pursue his interest and love of birds. He traveled globally to collect, but his great skill was in observation and writing of the details about birds. His foundational work was used by many publishing ornithologists of the time.

Bishop was a founding member of the NHBC and an early leader of committees, trips, and studies. He was a traditionalist, in that he felt ornithology should be centered on the collection of specimens. But he did join ranks with the club's younger generation, who favored birding with optics. The young founders —Saunders, the Pangburn brothers, and Honywill—looked up to him for guidance. However, in a letter to the club commemorating its thirtieth anniversary, Bishop reasserted his old-school beliefs, stating that he never truly supported the idea that birds could be studied without collecting them.

In 1918, Bishop moved permanently to Pasadena, California. He maintained a winter home in Hollywood, then largely undeveloped. He continued to expand his specimen collection, now mainly through purchases. Bishop published seventy letters in scientific journals, numerous articles, and several books. Among his publications were the NHBC's 1908 "List of Birds in the New Haven Region," the 1913 AOU Checklist, and many descriptions of birds and eggs from a wide range of locations. Bishop passed away on April 3, 1950, the club's anniversary.

Ernest F. Coe, another founding member, graduated from Yale and became a prominent landscape architect. In 1925, at the age of sixty, he moved to Florida and soon was involved in efforts to preserve the Everglades, becoming one of the strongest advocates for a national park designation. In 1997, the US Congress formally declared Coe as the driving force behind the park, noting that he is considered the "Father of the Everglades." Most people who enter Everglades National Park will stop at the east entrance visitor center named after Coe.

Middle Years

After its eventful first decade, the NHBC would settle into an efficient routine. Meetings were called to order by the president. A few business items were voted on by the general membership; then a discussion of recent bird sightings was followed by a presentation. Springtime bird sightings in the 1920s and '30s always mentioned the welcome arrival of the Baltimore Oriole. These birds built their pendulant basket nests in the stately elm trees throughout New Haven until the trees started to die off from Dutch elm disease in the 1940s. The sighting reports also chronicle the encroachment of the Northern Cardinal, Tufted Titmouse, and Red-bellied Woodpecker as well as the appearance and disappearance of noisy Evening Grosbeak flocks. The late Richard English, a past president and one of the club's longest-term members, carried the tradition of the bird sightings discussion into the first decade of the twenty-first century. But as the new century ushered in the use of internet-based group emails, international databases like eBird, and handheld devices that updated birders the second a rare bird sighting had been reported, the tradition became obsolete.

The bird sightings discussion and the business portion have been all but dropped from the club's monthly meetings. Business, except for officer elections and bylaw changes, was moved to a separate board meeting held monthly. (A common item of business was the purchase and distribution of wild bird seed to the members.) These changes left more time for announcements and the speaker. At first, the speakers would present lantern slides, then Kodachrome slides, and now computer-generated slide shows and programs. Some of the most popular and frequent speakers in the middle years of the twentieth century were Dr. Stanley Ball, a Yale associate professor and curator of zoology at the Peabody Museum; and the ornithologist and wildlife conservationist Dillon Ripley, who would become secretary of the Smithsonian Institution. Both frequently showed slides from their worldwide adventures.

Club membership waxed and waned in the 1930s and '40s, eventually starting to tick up to its current level of about five hundred members. The meetings continued to move around New Haven, usually taking place at Yale facilities. During World War II, meetings convened in the afternoon, due to the blackout orders that were enforced each night. The club purchased war bonds to support the war effort. The annual meeting and banquet were initially held each September; eventually they were moved to May.

From the club's earliest days, members went on field trips to favorite places throughout the county and the state. Local sites, some now forgotten, included Bishop Woods, Clintonville, and Clarks Pond. The club ventured as far as Rhode Island, Cape Cod, and Bombay Hook National Wildlife Refuge in Delaware.

In the 1930s, the club hosted a series of bridge parties, picnics, and banquets to raise funds to purchase a wooded plot of land of sizable acreage in an urban setting for a sanctuary. The club hoped to procure a forty-acre lot in the Montowese area and issued pleas for land donations as well as notices of fundraisers in the local papers. The nights of card playing, with crullers and birch beer for refreshments, raised some money, which was put into a designated sanctuary account, but it was never used to purchase any land. The fund was used to finance some of the club's expenses until, eventually, it was absorbed into the general account. The club engaged in social projects as well, organizing books at the New Haven Library, judging Boy Scout birdhouse-building competitions, and helping with art projects at area schools.

Bird banding continued to be a side activity for some members. An effort led by Mr. and Mrs. Harold Hutchins and Dr. and Mrs. Henry Bunting in May 1940 located two thousand Chimney Swifts roosting in a hospital chimney. Together the couples banded 750 birds. Three years later, in November, one of their birds along with a few others banded elsewhere were recovered in the Amazon tributary system in South America. This discovery finally answered the question about the wintering ground of the Chimney Swift.

Mrs. Hutchins and Mrs. Bunting, the busy bird banders, stand among the many women throughout the club's history—some known only by their husband's names—who have participated in and maintained the organization's activities. A number were educators. Lena J. Gorham and Maybelle Hayes, teachers in the New Haven school system, incorporated bird study in their curriculum. Lena Gorham would bequeath a sum of money to enable the club to bring in better speakers for the meetings. This fund eventually was combined with the sanctuary fund and then incorporated into the general fund.

In John's bookcase, there are far more books by men than by women. The fields of birding and ornithology have been dominated by men, although bird clubs across the country had women driving their activities. The NHBC has listed women as officers, chairs, and committee members since its earliest days.

The club began issuing a yearbook in 1917. Over the years, it has presented the club's activities, including indoor presentations and trips; its objectives, constitution, and bylaws; its officers and committee members; an "In Memoriam" section; and a list of past presidents. (Of the fifty-one past presidents listed in the 2020 yearbook, sixteen are women; four of the first ten were women.)

A drawing of a Least Flycatcher by the pioneering ornithological illustrator Louis Agassiz Fuertes ran on the yearbook's cover from 1936 to 1999. As the reproduction had diminished in clarity over the years, the flycatcher was retired in 1999, to be replaced, over the next three years, by images of a Swamp Sparrow, Prairie Warbler, and Belted Kingfisher. In 2003, the NHBC's Black-capped Chickadee logo was introduced and adorns the cover to this day.

The NHBC yearbook continues to be produced and printed as a booklet. It contains a list of members and the year's program schedule of lecturers, birding trips, and conservation and other events. John Triana's bookcase and the past yearbooks together give us a wonderful lens through which to see the individuals, unknown and famous, who have shaped the club, as well as a catalog of the activities and traditions that have kept it vital.

The New Haven Bird Club Today

In late September 2019, a small team of NHBC members showed up at the Great Meadow Unit of the Stewart B. McKinney National Wildlife Refuge in Stratford, Connecticut, to build an observation blind. The blind would overlook "Warehouse Pond," as it is known to birders. The team was very industrious, and the blind was completed hours before the planned finish time.

A member of the refuge staff looked on and stated under his breath, "We have been talking about putting this blind up for years now, and it goes up in just a few hours."

As the team members packed away tools in their trunks, a car pulled into the parking lot. Two birders jumped out and went to the pond and the new blind, its first users. After the hours of sawing, hammering, and drilling, there were no birds to be seen, and the birders soon returned to the lot. These birders had come down from the Hartford area to bird the coast, and the NHBC members were anxious to learn what they had to say about the new observation blind. Once the birders learned that the people soliciting their opinion were NHBC members, one exclaimed, "New Haven Bird Club? You guys are really busy. You have a lot of trips and lectures."

The club members enjoyed hearing that they have a reputation for being very busy. Over the past century, their activities have included indoor lectures, bird trips, conservation activities, education, hawk watches, supporting bird atlas projects, winter feeder watches, bird banding, breeding bird surveys, and other citizen science-based activities. These events have been open to the public at no charge, and people of all ages are encouraged to participate. It is a busy club indeed.

Hawk Watch

New Haven's Lighthouse Point Park is one of the premier sites in the northeastern United States for viewing the fall hawk migration. It was first recognized by the noted physicist and ornithologist Charles Trowbridge, who probably visited Lighthouse Point Park during the 1890s as part of his study of hawk migration in the New Haven area. It wasn't until the 1970s, however, that birders began formal documentation of the region's hawk totals. The history is recounted in this book in Steve Mayo's "Hawk Watch" and Arne Rosengren's "Dedication of the Hawk Watch

Plaque at Lighthouse Point Park." Today, NHBC members and others count the thousands of hawks and other raptors passing overhead, a passage that continues from the first autumn cold fronts, as early as mid- to late August, to late November or the first week of December.

The Lighthouse Point Park Hawk Watch site is bounded by Morris Creek and East Haven to the east, Long Island Sound to the south, and the New Haven Harbor to the west. The migrating hawks' flight direction is typically from east to west—that is, along the coast. Other migrants, mainly passerines, fly toward the northwest as they follow the edge of the New Haven Harbor. In extremely strong northerly winds, hawks often follow the Lighthouse Point Park beaches and then turn upwind to avoid the open water. These hawks may fly back over the Hawk Watch site as they get buffeted by the wind. On many days, hawks are seen flying low over the park during early morning hours. As winds slow, the hawks tend to fly at increasing heights, but they may be seen low again at the end of the day. Cold northwest winds from a passing cold front provide ideal conditions for hawk-watching. Strong west winds can also usher in large numbers. Light southerly winds or rainy easterly winds produce few hawks.

At many northeastern hawk-watch sites, peak flights coincide with the migration of the Broad-winged Hawk (*Buteo platypterus*). This species takes an inland flight route, peaking during the second and third weeks of September. But at coastal Lighthouse Point, large numbers are associated with the movement of *Accipiter* species and many other raptors during the second week of October. It is not unusual to tally many hundreds of individuals of thirteen species on early October days. The late season at Lighthouse brings migrating Turkey Vultures, Red-tailed Hawks, and Red-shouldered Hawks on bitter northerly November winds.

Hourly counts from nearly fifty years of hawk data have been submitted to the Hawk Migration Association of North America. Lighthouse was one of the first sites to become eligible to participate in the Raptor Population Index (RPI), one objective of which is to "produce statistically defensible indices of annual abundance and trends for each species of migratory raptor from as many count sites as possible." RPI data has shown the recovery of the Osprey in the Northeast, the success of the Bald Eagle, and the alarming decline of American Kestrels and, more recently, Sharp-shinned Hawks.

Hawks, as noted, are not the only Lighthouse Point migrants. Historically, Neotropical migrants (birds that breed in northern North America and winter in the American tropics) were known to pass through the site in September. These appear to be in decline overall, but impressive numbers of Ruby-throated Hummingbirds continue to pour through in September; thousands of Blue Jays, Black-capped Chickadees, Cedar Waxwings, Eastern Bluebirds, American Pipits, and American Robins follow in October; and hundreds of thousands of blackbirds can appear in a single November morning, rising like smoke from the east. Connecticut rarities seen over Lighthouse include Dickcissel, Red-headed Woodpecker, and winter finches.

The Lighthouse Point Hawk Watch, a highlight of the fall, provides a wonderful opportunity for new birders to learn from knowledgeable veterans. Since 2003, the NHBC has worked with the New Haven Parks Department to host the annual Migration Festival, held at Lighthouse Point in September. The club leads bird walks, provides hawk-watching expertise and educational information, and makes donations to sponsor various aspects of the festival.

The Big Sit!®

An annual, international birding event, The Big Sit! is a notably low-key, friendly competition, founded by the NHBC in

1992. The event has been coordinated since 2001 by *Bird Watcher's Digest*, a popular magazine with a target audience of birders and bird feeder enthusiasts. With the passing on of its publisher, Bill Thompson, the digest transferred management of the event back to the NHBC in 2020.

The Big Sit! occurs every year in mid-October. It is like a Big Day or bird-a-thon, but the object is to tally as many bird species as can be seen or heard within twenty-four hours from within a seventeen-foot-diameter circle. A circle can be organized for an individual or a team. All watches are completed within a two-day weekend. Participants place their circles at various vantage points, such as hilltops and observation platforms overlooking fields, wetlands, beaches, and other habitats, to optimize the chances of observing the most birds. The event has become known as a "tailgate party for birders," as elaborate snacks and beverages, along with the camaraderie and the friendly competition, have become an important tradition for many circles. A number of circles have clever names—Pishing in the Wind, the Boreal Birdometers, and the Surf Scopers, to name a few.

The Big Sit! is a high point on many birding organizations' calendars, including bird clubs, Audubon chapters, and friend associations of National Wildlife Refuges. Some groups use the event as a fundraiser, with circles soliciting pledges from sponsors. For many, however, the focus remains on enjoying The Big Sit! as a fun social event. Big Sit! circles have been registered in forty-nine US states, plus five Canadian provinces and twenty-seven other countries.

Connecticut Bird Atlas Projects

From 1982 to 1986, the birding community in Connecticut embarked on a project to compile an atlas of breeding birds in the state. The project was driven by the National Audubon Society, its local chapters in the state, the Connecticut Audubon

Society, and the Connecticut Ornithological Association. The NHBC provided financial contributions and volunteers. Club members assumed prominent roles as regional coordinators as well as authors and editors of the *Atlas of Breeding Birds of Connecticut*, edited by Louis R. Bevier and published in 1994.

Increasingly, state bird atlases are being redone, with a primary goal of determining whether and how bird distributions are changing. Although there is some communication and exchange of ideas between state atlas projects, they are independent efforts. A second Connecticut Bird Atlas project was kicked off in 2018, organized by the Connecticut Department of Energy and Environmental Protection and the University of Connecticut. The NHBC is once again a source of donations, coordinators, and volunteers.

The atlas project has divided the state up into grids based on US Geological Survey topographical maps. Each of these grids is divided into six blocks, yielding almost six hundred blocks across Connecticut. Birders, numbering over five hundred individuals statewide, adopt blocks to survey and then record the presence of birds and their breeding activity, generating lists of species and weighing the evidence that breeding occurs within the block. Breeding evidence is assessed by observing the birds and assigning their behaviors to predefined categories that represent possible, probable, and confirmed breeding. Species' wintering grounds are also recorded throughout the state. Seasonal employees are hired to conduct protocols to survey migration and species abundance.

NHBC members have contributed to the atlas beyond individual adopted blocks by "block busting." A group of eight to ten members splits up into teams, which spread out across a few of the survey blocks. By the end of the day, surveys of underreported blocks can be completed. The block-busting parties give members an opportunity to bird in areas they do not usually frequent, such as Pachaug State Forest in New London County and Mount Tom State Park in Litchfield County.

The first bird atlases that used science-based methods and analysis began in the United Kingdom in the 1960s. Many bird atlases have been produced since, focusing on areas from continents down to counties. Bird atlases provide species distribution and breeding data of immense scientific value. The follow-up atlases provide important data on population and phenology trends. Newer approaches using Geographic Information System (GIS) software tools are able to pin the species' occurrence and abundance data to habitat types. This information is valuable to conservation efforts, allowing for a focus on those habitats that host the most species richness or are essential to threatened species.

A Long History of Christmas Bird Counts

Well into its second century, the National Audubon Society's Christmas Bird Count (CBC) continues to be a very popular event. Founding members of the NHBC were participating in the CBC before the club was formed, and the club has participated in most Christmas counts ever since. The CBC has evolved a standardized protocol, mobilizing more than seventy-two thousand volunteer birders in more than twenty-five hundred locations across the Western Hemisphere. The count period extends from December 15 to January 5. Volunteers count the birds inside a designated circle with a diameter of fifteen miles in a twenty-four-hour period. With the number of volunteers and the longevity of the event, Audubon has been able to amass a large amount of data that show trends in bird populations, both alarming and encouraging. The data have been used to trigger wide-scale conservation efforts.

Every year, approximately fifty NHBC members cover the fifteen-mile circle centered on New Haven Green (latitude

41.3006920; longitude −72.9322830). The members brave cold, snow, rain, and wind to report compiled counts of, on average, 125 species. For many club members, participating in the New Haven CBC circle is not enough, so they branch out to support other circles in the state, in Guilford, Milford-Stratford, and Quinnipiac Valley.

The NHBC's first official CBC occurred in December 1907. Eight areas were surveyed by at least seven people over several days. Clifford Pangburn did not cover Edgewood Park, as he had the year before, instead choosing Lake Saltonstall. His brother Dwight, covering Edgewood and Mitchell's Hill on a chilly day of thirty degrees with very little snow on the ground, would count Purple Finches and Pine Siskins. In 1912, the Pangburn brothers reported a Ruffed Grouse and twenty-six Rusty Blackbirds, both first-time species for the New Haven Christmas count.

Rusty Blackbirds would continue to be counted in New Haven CBCs over the next 107 years. These observations, coupled with CBC data from throughout the country, indicate over time that the Rusty Blackbird is in trouble. Since 1966 the species has declined rapidly, by 89 percent. It is citizen science, which has been a cornerstone of the NHBC's activities, that has drawn attention to the problem. Ornithologists have now launched research projects to identify causes of the decline of the Rusties, so that conservation measures can be launched.

The Edgewood Park area has changed since NHBC members began surveying it for the CBC nearly 120 years ago. The Yale Bowl has occupied the area at the top of the park since 1914. Development of the area around the bowl in the 1910s consumed the old Mitchell and Greist estates as far west as Forest Road. The Yale Golf Course sprawls over Mitchell's Hill, and what was once a pond between the hill and Fountain Street was filled in to make space for a housing development and Brooklawn Circle.

This area is no longer the remaining untouched wild region of New Haven Colony that Clifford Pangburn entered during the CBC in 1906. The cliffs of West Rock still hang to the north, visible through the trees at certain vantage points. In addition to Edgewood Park, the old CBC sector now encompasses Marginal Drive alongside the West River, Evergreen Cemetery, the Yale Golf Course, and Beaver Ponds Park.

The New Haven CBC circle, like all the circles, can provide only a one-day snapshot of the birds and weather conditions in the December-early January time frame. The weather in southern New England has always been variable, but records show that temperatures rising into the forties and precipitation falling as rain were the exception on the CBC date. Recent trends over the past twenty years reveal that temperatures in the fifties and rain are becoming more common. Climate change may explain this trend in the weather as well as the sea-level rise in Long Island Sound, which impacts the salt marshes along the Connecticut shoreline. The National Audubon Society provides data and graphics to compilers to use in presentations that they may give to the public. One image the society provides depicts the northward movement of the wintering grounds of many birds. The trends revealed by the CBC may be explained by climate change, but there are many stochastic factors that obscure a clear understanding. The CBC serves as just one source of data.

What has remained constant through the century-plus is the type of NHBC member that has gone to survey the Edgewood area for the CBC. While many members are well educated (often affiliated with Yale, as faculty or alumni), all are persistent and intrepid. Like the Pangburn brothers and their friends, they have faced snow and cold, and they love birds. Dr. Bill Batsford, who has been covering the area since 2007, is no exception. A member since 2003, Dr. Batsford has served on the NHBC board as the Indoor Programs Chair, Conservation Chair, Vice President, President, and Outdoor Programs Chair. He actively birds the same hot spots that members have been birding since 1907. Like Herbert K. Job, he is seen frequently with a camera. In fact, many of his photographs appear on the club's website. He simply loves the birds. Dr. Batsford, like the Pangburns, is reporting about forty species every Christmas Count. During one CBC he counted 120 Rusties!

And More . . .

In 1937, from his home in Pasadena, Dr. Louis Bishop, probably in a nostalgic mood, wrote "History of Ornithological Studies in New Haven," to be read at the NHBC's thirtieth anniversary celebration. Bishop humbly plays down his role in the founding years, portraying himself as just an adviser to the boys who started the club. He also mentions the club's women, largely teachers in the New Haven schools. These young men and women, he insists, are responsible for the success of the club. He then lays out the real reason for the club's success over thirty years:

> No matter how fine the officers and how enthusiastic the members, a sustained interest in the subject [which] let no trouble interfere with their study was necessary those formative years, and this the members had. Devotion that will collect a crowd at 5am on spring mornings in the country to observe and study birds, and bring them week after week, is the kind of interest that counts, and they had that.

Although the membership has grown bigger and older over the decades, the club is still powered by the enthusiasm and devotion of men and women who are young at heart. Their interest still gets them up before the sun during migration and out into the snow and rain for a Christmas Bird Count. There is no way to know what the next century will bring in terms of changes in the landscape, the climate, the human environment, or the bird populations in New Haven County. The club is facing new challenges in 2020 with a pandemic and will reach out to new technologies for solutions to maintain its activities. But its members will not be daunted by an alarm going off at four o'clock in the morning, by the vagaries of weather, or by new challenges, as long as there is this enthusiasm, devotion, and interest.

The picture of Bishop that accompanies his obituary in the *Auk* shows a twinkle, a spark in his eye. This look embodies the very qualities he attributed to the young founding members. We have seen this light in the faces of schoolchildren participating in the club's educational programs at local schools through the decades. We see this light animating faces young and old at the sight of a Blackburnian Warbler on a field trip in May. The light is constantly renewed in new members who will lead the club into the future. It is this light, signaling enthusiasm for, devotion to, and interest in the birds that built the NHBC, carried it for a century and more, and will continue to do so.

Author's note: I wish to thank John Triana for his help in providing materials and direction for this project based on his many years of work doing research and compiling resources as the NHBC historian and archivist, and for his patience in fact checking and proofreading the chapter. I would also like to thank my wife, Jane Brokaw, for her support and her proofreading.

References

Barrow, Mark V., Jr. *A Passion for Birds: American Ornithology after Audubon.* Princeton, NJ: Princeton University Press, 1998.

Bevier, Louis R. *The Atlas of Breeding Birds of Connecticut.*

Hartford: Connecticut Department of Environmental Protection, Connecticut Geological and Natural History Survey, 1994.

"Bird Club on Bonnet Crusade." *New Haven Register*, May 3, 1907.

"Bird Club Plans for Field Work." *New Haven Register*, April 17, 1907.

Bird Lore: An Illustrated Bimonthly Magazine Devoted to the Study and Protection of Birds, 1900, 1903, 1904, 1906, 1907, 1909.

"Bird Lovers Here Organize a Club." *New Haven Register*, April 7, 1907.

"Bird Study Club Forms." *New Haven Chronicle*, April 6, 1907.

Burroughs, John. *Wake Robin* (1871), in *The Complete Works of John Burroughs*. Kindle Books, 2015.

Cole, Leon J. "The Early History of Bird Banding in America." *Wilson Bulletin*, June 1922.

"Honorary Members of the Bird Club." *New Haven Register*, April 7, 1907.

Howard, Hildegarde. "Louis Bennett Bishop, 1865–1950." *Auk*, October 1951.

"How Burroughs Became Member of Bird Club." *New Haven Register*, April 21, 1907.

New Haven Outdoors: A Guide to the City's Parks. New Haven: Citizens Park Council of Greater New Haven, 1990.

"Rapid Growth of Westville in a Few Years Almost without Parallel," *New Haven Register* (March 21, 1915), in *Westville: Tales from a Connecticut Hamlet*, edited by Colin M. Caplan. Charleston, SC: History Press, 2009.

"The Subtle Beauty of Nature," *Connecticut Magazine* (1904), in *Westville: Tales from a Connecticut Hamlet*, edited by Colin M. Caplan. Charleston, SC: History Press, 2009.

"To Form New Haven Bird Club." *New Haven Register*, March 26, 1907.

Van Der Aue, Kathy. "Mabel Osgood Wright and the History of Birdcraft & the Connecticut Audubon Society." Originally published in *Connecticut Warbler*. Connecticut Audubon Society, 2018, https://www.ctaudubon.org/2018/01/mabel-osgood-wright-and-the-history-of-birdcraft-the-connecticut-audubon-society/#sthash.RiuyUWLE.dpbs.

Triana, John. Letter to the Editor. *Connecticut Wildlife*, January/February 2007.

Weidensaul, Scott. *Of a Feather: A Brief History of American Birding*. New York: Harcourt, 2007.

Van Der Aue, Kathy. "Mabel Osgood Wright and the History of Birdcraft & the Connecticut Audubon Society." Originally published in *Connecticut Warbler*. Connecticut Audubon Society, 2018, https://www.ctaudubon.org/2018/01/mabel-osgood-wright-and-the-history-of-birdcraft-the-connecticut-audubon-society/#sthash.RiuyUWLE.dpbs.

Triana, John. Letter to the Editor. *Connecticut Wildlife*, January/February 2007.

Weidensaul, Scott. *Of a Feather: A Brief History of American Birding*. New York: Harcourt, 2007.

PART II

Birder Beginnings

Why We Bird

Gail Martino

May is a rejuvenating month in New England. If you open your window on a warm May morning, even on a residential street in New Haven it can sound as though you are in the midst of a wildlife preserve. If you take a moment to detach from your daily routine—close your eyes and listen—you might hear a Carolina Wren belting out its song, Mourning Doves cooing backup vocals, a Red-bellied Woodpecker drumming a backbeat, and Northern Cardinals cheering on the chorus. For a birder, these are moments when worries of the day recede and connection to the wider world comes to the forefront. Essentially, it is when the ordinary becomes extraordinary.

As I sit at my desk writing this essay in the middle of a New England winter, reminiscing about these extraordinary moments, I feel impatient because I must wait for the beginning of spring to experience this annual rapture. However, I did not always feel this way. In early childhood, I was curious only about the flight of the birds passing high above. When I was nine years old, my father helped expand that curiosity. He showed me books containing photographs of the various birds and taught me to differentiate the birds in our backyard. I became fascinated by the behavior of the birds around me.

This shared experience with my father led to my decades-long passion for birds and birding. Like me, all birders, it seems, have a personal "trigger" story of how they initially became interested in birds and birding.

Since a key goal for any bird club or group is to encourage an interest in birds, birding, and the natural world, it is a good idea for birders to develop an awareness of their trigger stories. "How did you get interested in birds?" is a question that many of us birders have been asked by non-birders. Our readiness to recount our trigger stories can help spark, foster, and/or increase others' appreciation for birds. To this end, the New Haven Bird Club (NHBC) sought to assemble the trigger stories of a broad range of birders. Some of these stories turned into essays in their own right, which appear elsewhere in this collection. The essay "The Gifts of Richard English," for example, grew out of anecdotes about a late club member that were submitted independently by members influenced by him. One member's discussion with another is a separate piece titled "Interview with Arne Rosengren."

We sifted through the trigger stories members submitted to identify connections and commonalities in the formative experiences that prompted an interest in birds. This essay explores what we learned by collecting and analyzing these stories and their themes.

Trigger Stories: Common Themes

Do you have a birdwatching trigger story of your own? In response to our request for stories, a diverse set of NHBC members—birding novices and experts, men and women, young and old—shared personal experiences that precipitated their enthusiasm for birds.

The respondents were generous with details—sharing personal backgrounds, birthplaces, early experiences observing birds, related reflections, and contributions to birding. Their stories provided glimpses of the purpose birding has served in their lives. A common theme among the respondents is a practice of communicating their love of birds to friends, family, or their community. For some, birding is central to their lives, both vocation and avocation. For others, it is a hobby enjoyed during free time. Notably, some respondents described a shift from occasional birding to a more focused pastime subsequent to a specific change in lifestyle, such as retirement from full-time work. Beverly Propen told us, "It wasn't until after my kids graduated college and left the house that I found time to pursue my interest in birds and birding."

Formative Experiences in Childhood

Respondents who shared a trigger story most often mentioned direct or indirect childhood experiences they believed were fundamental to their interest in birdwatching. Some members credited exposure to a particular environment—such as living in a wooded area in childhood, whether in Connecticut, New York, New Hampshire, or Canada—for its impact on their lifelong appreciation of nature. Gilles Carter said, "Growing up in Canada, I was always interested in the outdoors and venturing out in nature. The family and I would often go out when it was thirty to forty degrees Fahrenheit to appreciate nature. I've always found severe conditions challenging and interesting." Beverly Propen grounded her childhood experiences in growing up with a pretty backyard in Flushing, New York.

Sometimes a chance observation served as trigger. For the NHBC's First Wednesday Walks coordinator, Tina Green, it was a picture of a Cedar Waxwing that caught her attention as a second or third grader and ignited her passion for birds. She was captivated by the intricate detail of the red tips of the secondary wing feathers and the overall sculptural beauty of the bird itself, with its black mask and prominent crest. In the years following that encounter, Tina contented herself with studying artwork associated with birds in various iterations—paintings, photographs, sculpture, and other representations. And as a young mother, she spent many happy hours with her children at the Yale Peabody Museum of Natural History in New Haven, which displayed an outstanding bird collection.

The Impact of Promoters

Nearly all NHBC members credited a "promoter" (a mentor or guide) who helped foster their interest in birding at a key point in their lives. These promoters ranged from a family member, friend, or mate to a walk leader or other skilled birder. Birding interest was sparked by something as simple as a bird-loving person (a promoter) hanging up a bird feeder or asking someone to go on a birdwatching walk.

Some responses revealed that promoters are important for both children and adults and that the "triggering" can at times be gradual rather than a single event. Beverly Propen reflected, "I have always loved animals (my brother is a vet), so my parents must have instilled in us a love and respect for wildlife." Valerie Milewski, a member whose interest in birding was awakened as an adult, said playfully, "It's my friend Laurie's fault for bringing me along on bird walks!" Jack Swatt said: "At one point, I volunteered at the Shepaug Dam [a Bald Eagle observation site in Southbury, Connecticut] after seeing an advertisement. Connecting with other birders there encouraged me to learn more about birds. Dave Rosgen [a friend] was a big influence."

Alan Malina shared a story of how a family friend had a powerful influence on his and his wife's interest in birds:

> A nice spring day in 2001, my wife and I were walking on the Farmington Canal linear path in Hamden. We heard

singing that we could not help but stop and admire, and we tried to see what the source was. Finally, we saw a beautiful black-and-white bird with red on its breast. When we returned home, my wife called her mother's best friend, who she knew was a birdwatcher. It was a Rose-breasted Grosbeak. About a week later, my wife received an unexpected package from this friend with a *Peterson's Field Guide to Eastern Birds* and a pair of binoculars. Thus, a hobby was born!

Close Encounter Stories

For some members, experiencing a "close encounter" with a particular bird had a lasting and/or amplifying impact on their overall interest in birdwatching. Beverly Propen described an opportunity to care for a bird when she was just a child in the 1960's, after a neighbor had dislodged a nest. At the time, she did not realize how disruptive dislodging a nest was to the bird's life cycle. However, it did make a lasting impression, as she reveals:

> When I was about twelve years old, my neighbor took down a bird nest that was in his air conditioner. There were three baby starlings in that nest. He kept one, gave one to another neighbor, and gave another starling to me. I kept it in a shoebox with tissues inside and with holes for breathing. I fed it raw hamburger and bologna. The starling grew and flourished. It would perch on my index finger, and if I would remove my finger, it would flap its wings and start to fly. I don't remember how long I kept it, but one morning when I opened the box, the starling flew up to a tree in my yard, and that was it. At the time, I was so happy because I figured it was on its own. I realize now that a young bird still needs its parents' care for a while after it fledges. But I have sweet childhood memories of that starling.

Shifting from Looking at Birds to Seeing Them

Many respondents noted a shift from looking at birds to seeing them—and this further sparked their interest in acquiring more knowledge about birds and birding. For them, birds were no longer generic living creatures; they became detailed and distinctive—ultimately evoking an emotional response to their activities.

One of the best examples of this perception change was articulated by Glen Cummings, who described how his formative experiences gradually helped focus his perspective on birds:

> There were several times in my life when I became interested in birds. When I was a child, my grandparents would place feeders in the yard, and I would watch the Blue Jays bully the other birds. That's the first time I realized you could watch them do something. Then, in the fifth grade, I camped in the woods by myself, and a Great Horned Owl landed in a tree not far from me, and I really saw how beautiful it was. I watched for a long time. It was an incredible creature that I remember to this day. In my twenties, I happened to start a conversation with a guy at a bar and told him about a strange bird I had seen. Based on my description, he said it was most likely an American Bittern. I was so impressed that it was possible to know the differences among birds from just a description. I bought a field guide to learn more about the differences among birds. I wasn't looking anymore, I was seeing.

Glen's wife, Ricci, the co-compiler of this anthology, described how he helped her develop a new appreciation of birds. When they were first dating, they did some incidental birdwatching while on a vacation.

"Hey, look! It's a Rufus-sided Towhee!" (now referred to as the Eastern Towhee), her future husband exclaimed.

Ricci responded, "That's ridiculous, there's no bird by that name!"

Her future husband begged to differ. She later looked it up in a bird guide and discovered his identification was correct.

"Prior to that experience, all birds were just generic to me. I didn't see the distinctions. After that, I got more interested in birds, and in HIM!"

Seeing to Learning

Once distinctions among different species of birds are appreciated, many birdwatchers develop an interest in learning more about the birds themselves. Valerie Milewski, a relatively new NHBC member, provided a good example of this shift toward a desire for increased knowledge. She shared her excitement about seeing new birds and how this encouraged her to learn more about them:

> In spring 2017, I was on a walk with Frank Gallo at East Rock Park in New Haven. On that day we saw over one hundred Yellow-rumped Warblers. It was amazing! This past winter I was on a walk with Frank Mantlik at Milford Point. I saw my first Snowy Owl on that walk. I was so excited about it. Frank was too. It was a really great find. The more I see the more I want to learn on my own and the more excited I get about birding.

Another member told us how she went about acquiring the increased knowledge she wished for. After she retired and had more free time, she attended a course at the Connecticut Audubon Coastal Center at Milford Point—and this course taught her how to observe more detail when learning to identify a bird.

One vital aspect of such learning involves understanding where certain types of birds live and then trying to view them in that habitat. It can later mean learning the songs of these birds. Some members mentioned first learning about the birds in their own neighborhoods—and then acquiring knowledge of birds in other locales.

Jack Swatt talked about getting more involved in walks over time as he learned more about birds and their characteristic songs. Along these same lines, Gilles Carter, after a chance encounter with an ornithologist at Yale University, took the professor's advice and enjoyed his first birding vacation in Arizona, where he was "blown away" by the hummingbirds he saw—all new to his life list.

Learning to Engaging

Many NHBC members remarked that once they began to learn about birds, they decided to become involved in citizen-science projects and related activities. An example of such engagement was shared by Jack Swatt, a member who monitors Eastern Whip-poor-wills in Naugatuck State Forest, north of New Haven. Jack first heard the whip-poor-will's onomatopoeic bird call on a July Fourth many years ago, just before the Independence Day fireworks display at Southington Mountain. Not long afterward, he read a disturbing report that Connecticut sightings of Eastern Whip-poor-wills were declining. When the Connecticut Department of Energy and Environmental Protection (DEEP) advertised for volunteers to survey the birds in a geographic locale close to his home, Jack jumped at the chance. At first, a Connecticut DEEP survey required solely a drive-by, to listen for the birds. Later, it involved venturing into the woods in the evenings. As he explains, "I preferred to survey at two or three in the morning, because it was easier to hear the birds when there was less car noise. In Naugatuck State Forest, I have heard as many as five males calling in an evening." For over ten years, he has conducted these surveys on moonlit evenings (a key cue the birds use to start calling). Moreover, he continued his monitoring of birds even after the official bird study ended.

Many other respondents reported that they had likewise become involved in citizen-science projects or other endeavors as a way to become more informed about birds and engaged in their well-being. One member explains how she became involved in bird rehabilitation:

> I found a mockingbird with an eye injury on the walkway from my apartment. I picked it up, placed it in a box, and looked up the contact information of a wildlife rescue organization. I drove the mockingbird to the organization. The owner said the mockingbird would probably be blind in the injured eye, and therefore it could not be released back into the wild. I ended up volunteering at that rescue organization on my days off from work.

Gilles Carter, a videographer by training, described how he engages with the birding community through Connecticut Audubon film projects. In 2018, he filmed scenes in East Rock Park, a local Important Bird Area. That film highlighted mating, nesting, eggs, feeding, fledging, and migrational departure of birds—a full year's cycle of birding—in this New Haven park.

"Birds are so fun to film—they are real eye-candy!" he said.

Engaging to Sharing

Some NHBC members with more birding experience and knowledge explained that they are now focused on actively sharing their birding passion with others—in effect, they have become promoters or mentors. Jack Swatt, who once attended walks in order to learn, now leads walks in Naugatuck State Forest each June to introduce others to the Eastern Whip-poor-will's distinctive song.

The motivation to help others become excited about birding is illustrated by this story from Gilles Carter:

> About five years ago, my mother suffered a stroke that limited her mobility and prevented her from enjoying the birds in her garden as she used to do. Since then, I started

showing her photos of the birds I take in the field. Despite the changes to her medical status, sharing these photos with my mother has helped her connect with something she really enjoys and helps me stay connected with her.

Gilles also described how his films can reach larger numbers of potential birders and enable them to appreciate these creatures by revealing insights about avian behavior that can be gleaned through careful observation over time:

> For me, [birding] is not about the numbers of birds I see. It's all about the narrative unfolding. So, it's just as valuable to watch a common bird as an uncommon one. For example, I filmed an American Robin and noticed how, by presenting a worm, she seemed to lure her chicks away from me as they hopped on the ground. It seemed that she was showing the chicks that they should eat the worm and stay away from potentially dangerous videographers. Another time, I filmed sapsuckers for two weeks. During this time, I saw them set up tap holes and return to them in eighteen-minute intervals. It turns out sap takes time to run, so the woodpeckers would make a circuit, hollowing out one hole at a time, only to fly back nearly twenty minutes later and feed. While [the sapsucker was] away from the hole, I saw Yellow-rumped Warblers and later Red Admiral butterflies taking advantage of the liquid glucose meal incidentally provided by the sapsucker. Being able to share these stories of bird behavior to a wide audience through film is important to me.

Come for the Birds, Stay for the People

While all NHBC respondents were interested in birds, they also mentioned the importance of the human connections they have built through their experiences with the club. In turn, these experiences have helped them keep active and engaged in the birding community. As one member told us: "I joined the New Haven Bird Club so I could go on more walks. I liked getting together with others, the social aspect. A day when we see common things is just as good as when we see something unusual."

Sharing the Passion

As members talked about the various formative experiences that may have led or predisposed them to develop an interest in birds, a common outcome became clear. Many found that their interest grew over time, awakening a desire to share their passion for birds with the larger community of both birders and non-birders, creating a cycle of engagement.

It is never too late for you to start your journey as a birder. If you are not yet a member of the NHBC or another birding club, joining can be a terrific way to begin cultivating your passion for birds. On the other hand, if you are already an avid birder, joining the NHBC or another birding club can enable you to promote birds and birding and, hopefully, encourage future generations of bird-lovers and birders.

Editors' note: The quotations by respondents in this essay have been edited for clarity or brevity, but the key points provided by the respondents were preserved. Thanks to all members who shared their stories with us.

Interview with Arne Rosengren

Florence McBride and DeWitt Allen

This conversation was recorded on February 2, 2020, between Arne Rosengren, a 48-year member of the New Haven Bird Club, and his friend Florence McBride. DeWitt Allen provided the transcription.

FM: *How did you get interested in birds and birding?*

AR: Our summer cottage had House Wrens nesting on the porch. I slept out on a hammock, on the porch, and in the morning, right at dawn, they'd wake me up with constant singing and feeding the little ones, right over my head. And we had lots of beautiful Mountain Laurel bushes right in front of our cottage, forty or fifty feet down to the lake itself. I'd go in there while [the parent bird] was out looking for food. I'd sneak out and a couple of the nests were low enough I could see in there, see the eggs. My father tried to get me interested in fishing in the lake, and I tried, but nothing ever happened, just these little sunfish, and I thought it was boring. Boring.

FM: *How did you come to join the bird club?*

AR: It was pure accident. It came up with someone I used to go hiking with, something about birds, and he said, "Oh, I know someone who knows about birds. He belongs to the New

Arne Rosengren. Photo Florence McBride.

Haven Bird Club." I didn't know there was one. The friend said call up so-and-so, who's the new president of the bird club, and I called him up. He invited me to come to the next meeting. I was amazed how many people were there. Fifty or a hundred. They asked me to join the club, and I did immediately. Every two or three weeks they had somebody guiding a group; they still do it. So, I immediately joined up, and that took care of me. I was hooked.

FM: *About how old were you then?*

AR: I was a full adult, a middle-aged adult. This was thirty or forty years ago, [whenever] it was. There was a couple, George and Millie [Letis]; they were very helpful to me. They became very good friends of mine, besides just being birders. George was good at organizing things, meetings and so forth, so he was valuable to the club, even though he was more of a social birder. George and Millie, and Dick English was very active, and I hooked up with him, and I went birding with English a lot. I learned a lot from him, obviously, and I went on the club trips, and I learned from that.

FM: *Do you remember any birds you especially liked or bird trips that were special?*

AR: I was the one that started the daily Hawk Watch at Lighthouse Park; there was never anyone who was doing it daily. Pretty soon I had another person or two joining me. I'd only spend half a day or so down there at Lighthouse Point, but after a year or so of the bird counting, there were other people who were interested in what was going on with bird migration, especially hawk migration. We were getting twenty or thirty thousand hawks a year coming through there. Soon Ed Shove from North Haven started joining me down there, which was good because we had company, and he would cover the afternoons. We had so many hawks coming through that people were skeptical; they thought I was just padding it to make it look good.

Anyhow, that was a good day when I heard about the New Haven Bird Club from this man that I'd never met. He invited me to come down to the meeting, and I've been going ever since. I had an extra bonus out of the whole thing. I not only learned about the birds, where to go and how to see them, how to identify them, but I made a lot of very important friends, which I hadn't anticipated when I joined the club. So, it all worked out you know. Happy day.

PART III
Special Programs

Ed's Count
Frank Gallo

The Christmas Bird Count, the focus of this fictionalized story, is one of the New Haven Bird Club's longest-running events. This essay is dedicated to the memory of Ed Shove (1913-1994), a consummate naturalist, and a friend. He saw more than most.

It's six a.m., and Ed Shove is here to do a bird count. Before him stretches New Haven's Quinnipiac Marsh, his count area for far longer than he'd bother to remember. He walks quietly along the railroad trestle at first light. There is no need to come earlier; the owls will still be there—or will they? A mega movie cinema now replaces his best owling area. There have been changes, deemed "progress," over the years. He wonders if the owls see it that way. Behind him, the Middletown Avenue landfill stands in silent silhouette against the morning sky, a reminder of human presence and our effect upon the land. He moves on with a purpose.

Across the marsh, holiday lights from the houses along State Street twinkle in the clear morning air. It is a week until Christmas. He checks his thermometer. It is five degrees Fahrenheit.

It is the day for members of the New Haven Bird Club to participate in the National Audubon Society's annual Christmas

Bird Count. There should be forty to fifty species to reward him for his day's work. What will he find this year? There is not much unfrozen still water, so ducks should be in the river, if the hunters have not pushed them out. As he walks, he looks and listens for his quarry. The marsh is a harsh environment in winter. If you are a small bird or mammal, it is wise not to remain in the open. The wind is cruel, and the hawks are hungry. To his right, a Northern Cardinal calls. He marks it down. There are always the regulars: flickers, crows, starlings, and titmice are all here. Perhaps there will be a Golden-crowned Kinglet or a Hairy Woodpecker this year. Along the tracks, Northern Cardinals, White-throated Sparrows, Black-capped Chickadees, Northern Juncos, Tree and Song Sparrows all feed. There will be other, less common birds, such as the occasional Swamp Sparrow, Sharp-shinned or Cooper's Hawks, or perhaps a Brown Creeper. Then, there are always a few "good birds," like the Blue-winged Teals in 1979 and '80 or the Boreal Chickadees in 1977 and '83. Maybe there will be another American Bittern, like last year. They are out here; they're just so hard to find.

A startled Great Blue Heron flushes from a hidden river channel with a loud squawk; two steps later, a Black-crowned Night-Heron does the same. Both are uncommon birds for this time of year. A startled birder records them in his notebook. Farther down the line, a female pheasant rushes across the tracks and disappears instantly into the grass. Like the bittern, it is a difficult bird to find. There was only one here last year. A flock of seven Pine Siskins and two Common Redpolls alights in a nearby tree. It looks as if it may be a good "winter finch" year. A Northern Mockingbird jumps up to eye him from the top of a nearby rosebush, and two American Goldfinches pass overhead uttering their *Po-ta-to-chip* call. He writes them down, checks his watch and thermometer, and moves on. It is eight-thirty a.m. The temperature has reached fifteen degrees.

He follows the tracks deeper into the marsh. The tracks offer an unobstructed view of the marsh but no respite from the wind, which blows now from the north. All he can do is hike up his collar and press on. People are counting on him to find Rough-legged Hawks and Northern Saw-whet Owls. There are usually a couple of each wintering in the marsh. Looking up, he sees the familiar shape of a Red-tailed Hawk soaring over the dump. Beneath it, a Northern Harrier glides effortlessly on raised wings, hunting along a mosquito ditch. Another large hawk quarters the marsh in the distance, but he can't quite make it out. It begins to hover in place. It is a light-phase Rough-legged Hawk. As it nears, he can see its white rump and dark belly. That's one. Perhaps the dark-phase bird seen last year has also returned.

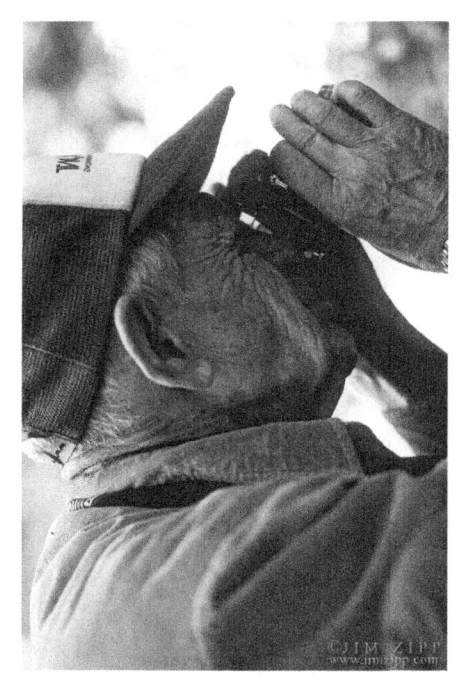

Ed Shove looking though binoculars. Photo Jim Zipp.

Before the day is through, he will return to the dump to count the many gulls. Maybe a Glaucous or Iceland Gull, rare winter visitors from the north, will be present to reward him for his work. Below the Middletown Avenue bridge, there should be Canvasbacks, American Black Ducks, and Common Goldeneyes, and with luck a Hooded Merganser or Pied-billed Grebe. If he has time, he'll check the bushes and marsh below the bridge for songbirds before checking the cedars off Sackett Point Road for saw-whets. The Eastern Meadowlarks that were near the bridge in late October may still be around.

A Killdeer calls from somewhere to his left. Searching, he sees a Belted Kingfisher plunge into the river and come up with a small fish. So far, it has been a good day. He writes these sightings in his notebook, checks his watch and thermometer, and moves on.

In the distance, he hears gunshots; he'll have to hurry. A flock of ten Canada Geese comes hurtling down the river away from the sound, followed by a group of Mallards and American Black Ducks. Wait—is there a smaller duck with them? He raises his binoculars. Yes, the smaller bird is a Green-winged Teal, a "good bird" for the marsh. It's the first since 1982.

He hurries on. His luck may not last forever. Another gust of wind greets him as he rounds the bend. He can see the river, and there are still ducks on the water. His eyes water as he tries to make them out. Ah, if only it were spring. The rails and bitterns would be back. He could watch the Least Terns in comfort as they fly up and down the river to feed, or spend leisurely hours poring through the flocks of shorebirds that frequent the marsh pools and riverbanks. There is no use lamenting his lot. It is not his way, and in truth he loves all this. Spring will arrive in due course. For now, people are counting on him, and he has ducks to count.

His name is Ed Shove, and he is here to do a bird count.

Hawk Watch

Steve Mayo

Origins of the Hawk Watch can be traced to when Neil Currie, a founder of the New England (now NorthEast) Hawk Watch, bumped into George Letis down at Lighthouse Point Park in New Haven one September day in 1970. Of course, either would head toward anyone holding binoculars, like moths to flame. And both were committed to introducing strangers to a lifetime passion for birding. Neil later ran into Arne Rosengren at Lighthouse. By 1974, Arne and Quinnipiac College undergrad Margie Pitcher had begun spending multiple days counting hawks during peak migration in September and October. Full-time coverage had begun. It became apparent that Lighthouse was special, and coverage grew. Ed Shove, Sal Masotta, Tony Tortora, and other NHBC members joined Arne at the watch.

Ed Shove had finally retired "from the mill" by 1980 and spent virtually every day of every autumn at Lighthouse Point Park tallying hawks. He was compiler until his death in 1994. Ed would begin his day by driving down Quinnipiac Avenue from his North Haven home and entering Lighthouse Point predawn. He'd walk the park in search of owls, often joined by New Haven park ranger Dan Barvir or Frank Gallo or Tom Mason. Others, many of them retirees and visitors, would then join Ed at the watch site in the parking lot. If things were slow, they'd stand around and talk football and eat the apples Ed had brought from Bishop's Orchards. But things were rarely slow, even with onshore winds. It was fall, and the hawks had to move. Accipiters poured through, and there would be yells of "the gang's up" as a ball of starlings over the marsh would escort a Sharp-shinned Hawk. Ed would scribble hashmarks, often giving up and writing 10s, 15s, and 20s in his notebook. At night he'd sit at his kitchen table and transcribe with pencil (and eraser) those hundreds of entries. They went onto a "green sheet," the Hawk Watch reporting form. There was no Excel back then, and the hourly totals and species totals had to match up exactly at the bottom right corner of the form. Bleary-eyed, Ed would hit the park again, for hours, the very next day.

On weekends, Jim Zipp was at his bird-banding station down in a mowed phragmites field a few steps from the Hawk Watch. He'd regularly come up with banded hawks for the crowd to admire. Holding a tiny male Sharpie and an enormous female Coop, he'd say, in mock disbelief, "How can you guys possibly get these confused?" Later in the season, compilers from all over New England would come south and visit Ed at Lighthouse Point. Their season, the inland migration of the Broad-winged Hawk, was over, and it was time to see American Kestrels, accipiters, and later buteos, low over the park, on stiff north winds. Old friends would be invited on November afternoons to join Ed and drive up to Banton Street, Valley Service Road, the Lucy Hammer estate, or some other nearby daytime owl roost. There was also the occasional national birder, with 599 or 699 species on his national American Birding Association list, who would telephone Ed for a guaranteed Northern Saw-whet Owl in a cedar just six feet away.

Cooper's Hawk at Lighthouse Point Park. Photo Abby Sesselberg.

While hawk-watchers were ticking twenty to thirty thousand hawks per season, passerines were pouring through in the early morning hours. The phenomenon was known by John Cameron Yrizarry, Noble Proctor, Richard English, and others, but there was no formal attempt at seasonal counting. John Granton made some preliminary counts in the 1980s, but most of the old-timers just considered these birds "little brown jobs." Besides, with all the hawks, there was no time to count the other birds. Systematic attempts to quantitate these migrants began with Greg Hanisek, who had moved to Connecticut from New Jersey to take a job at the Waterbury *Republican-American* newspaper. It would be amazing to see any data from those old days!

Ron Bell took over as hawk compiler, stationed on the "hill" in the center of the Lighthouse Point Park lawn. Tree growth had forced this move from the small parking lot watch site to the east. An active member of the NorthEast Hawk Watch organization, Ron made significant contributions to the analysis of the Hawk Watch data. He recruited many hawk-watchers who shared his love of the craft and was able to find someone to cover each day of the week. Some of these watchers are still holding the pen on their assigned day, a quarter century later.

To this day, every autumn, hundreds of people watch the weather and await a frontal passage with strong northerly winds, and some may even call in sick to work. Under the right conditions, they are guaranteed a big hawk day and the potential for a mega rarity. Several hundred hawks of twelve or thirteen species are still a common occurrence in early October.

Lighthouse Point Park has become a favorite among photographers, because hawks are lower—not the specks in the sky typically seen at other New England hawk watches. Lighting conditions during the afternoon "falcon follies"—when Merlins dart about the park—are particularly good.

Rarities recorded at Lighthouse have included Tropical Kingbird (1990), Calliope Hummingbird (2006), and Zone-tailed Hawk (2016). The latter occurred at the September Lighthouse Point Park Migration Festival, an annual event since 2003. That dihedral wing posture and striped tail sent people scattering from the watch site throughout the park screaming "Zone-tailed Hawk, Zone-tailed Hawk!" much to the surprise of casual visitors and the delight of birders.

Quantity, as well as quality of birds, is another hallmark of Lighthouse: three Short-eared Owls heading south, mid-morning; eleven Sandhill Cranes heading west over the oaks; scattered Dickcissels every September; Cave Swallows, more than a hundred, after starting in Texas and probably finding their way to the Maritimes, now swirling over the park; hundreds of Bobolinks; over 450 Ruby-throated Hummingbirds on a September day; thousands of Blue Jays against the orange and brown oaks; forty-five hundred Purple Finches, giving their *dit-dit* calls, overhead on an October day; and on November mornings, tens of thousands and sometimes hundreds of thousands of Common Grackles, like miles of smoke rising over East Haven.

Sheer numbers aside, the briefest of migrating bird experiences can be unforgettable. We've seen a Merlin tearing the hackles from a Golden Eagle; a hummingbird repeatedly diving on a Peregrine Falcon perched on the lighthouse; a Northern Wheatear bobbing up and down while perched atop a nearby pole; Lincoln's Sparrows lurking in the butterfly garden; Brown Pelicans coursing over the harbor; sluggish cuckoos in the nearby woods; a Cooper's Hawk chasing a squirrel around a tree trunk; American Woodcock flushed from the knotweed; Cattle Egrets and snipe directly overhead; a Northern Goshawk darting between us right at the watch site; eiders and Northern Gannets resting on the Long Island Sound; a Barn Owl teed up on a sign at the end of the day; Snow Geese like tiny white beads against a blue sky but still heard calling; Cooper's Hawks chasing Red-headed Woodpeckers; a Merlin capturing and tearing into a Tufted Titmouse a hundred yards away, its feathers almost immediately landing at our feet; Eastern Meadowlarks on the lawn; a Barred Owl perched on a Staghorn Sumac, unaware that the morning fog had dissipated.

Bird populations may be declining, but Lighthouse Point Park continues to be perhaps the best place to enjoy the spectacle of the Connecticut fall migration, learn hawk identification, and meet birders with like interests.

Mega Bowl

Chris Loscalzo

The cold winds lash our faces. The sea churns and smashes the shore, icing over boulders as the saltwater freezes when it meets the frigid air. We are dressed as if for an Arctic expedition, wearing layers of clothing, down jackets, fleece hats, and heated gloves. Why are we outdoors on this cold midwinter day? We are participating in the annual Mega Bowl of Birding, of course!

The Mega Bowl of Birding in New Haven County is a relatively new event for the Connecticut birding community, founded in 2017. Held annually on the first weekend in February, this is a friendly competition, in which birdwatchers, in groups of three or four people, go anywhere they like in the county to see birds. The teams keep a record of their observations and earn points for every species they see, from one point for the most common species to seven points for the rarest. Teams earn additional points by reporting rarities to the Mega Bowl coordinator, who relays the information to the other teams. The participants create whimsical names for their teams, such as the Tuff Ducks, the Not-so-Great Egrets, and the Winter Wrenegades. All the participants meet at the end of the day for a celebratory dinner, where they share stories and enjoy fine food. Every participant receives a prize, and special prizes are awarded to the members of the team that garnered the most points, as well as to the youngest and oldest participants. The winning team's name is inscribed on the Mega Bowl trophy.

The Tuff Ducks team braving the elements for the Mega Bowl in 2020. From left to right: Emmeline Kaiser, Bill Batsford, and Christine Howe. Photo Christine Howe.

The Mega Bowl celebrates birds and birders: Birds are beautiful, resourceful, adaptable, and resilient. Birders are intelligent, sociable, determined, and conservation-minded. Birdwatching has been a popular pastime for more than a century. The Mega Bowl affords local birders an entertaining way to look for birds and share their appreciation of them with one another. The competition is held in midwinter because at that time there is no migratory movement of birds, and there's not much movement of birders either. The absence of migratory activity makes it easier to predict the prevalence of the birds on the day of the event and designate the point value for recording each species.

We're out at the far reaches of Hammonasset Beach State Park in Madison, scanning Long Island Sound. At first, we see few birds, as the choppy sea obscures our view. Then we start to find them. We see Surf Scoters, Common Loons, and Long-tailed Ducks in the open water. We see Ring-billed and Herring Gulls flying fast, the wind at their backs. And we see dozens of shorebirds hunkering among the rocks and then lifting into the air, flying in unison as they look for a place to roost or feed. We see Dunlins and Ruddy Turnstones and a few rarer species as well: a Black-bellied Plover and a Purple Sandpiper hide within the flock. Watching these amazing creatures endure the elements buoys our spirits and helps us brave the chilly air.

The points awarded for the sighting of each species are determined by a number of factors. These include a species' relative abundance in midwinter, its distribution within the various habitats in the county, and its elusiveness. Some species, such as the group of irruptive birds known as winter finches, will vary in abundance from year to year, as their movements are affected by the annual cone crops in the northern coniferous forests. When cone seeds are bountiful, the birds stay in the boreal forests and are rare in Connecticut. In years when the seeds are scarce, the finches will travel down to our area in search of food. This variability prompts a change in their point value from one year to the next.

Species whose sightings are awarded one point are common and often found in large numbers. Two-point birds are fairly common and can usually be found with a little effort or luck. Three-point birds are uncommon and might be found only in specific locations. Four-point birds are uncommon to rare and are seldom found in the county in midwinter. Five-point birds are very rare and might be found in New Haven County only once every decade or more. Seven points are awarded for species never recorded previously in New Haven County in early February. There are about a hundred species that earn the observers one, two, or three points, and another hundred species that earn them four or five points if sighted. That there are as many rare species to be pursued as there are common ones makes the event exciting and unpredictable.

Examples of one-point birds include Canada Goose, Red-tailed Hawk, Herring Gull, Downy Woodpecker, and White-throated Sparrow. Among the two-point birds are Brant, Red-shouldered Hawk, Hairy Woodpecker, Carolina Wren, and

American Tree Sparrow. Three points are awarded for sightings of Snow Goose, Bald Eagle, Pileated Woodpecker, Cedar Waxwing, and Field Sparrow, among other species. Examples of four-point birds are Cackling Goose, Rough-legged Hawk, Yellow-bellied Sapsucker, Brown Thrasher, and White-crowned Sparrow. Rarities that garner five points for their discovery include Tundra Swan, Golden Eagle, Black-headed Gull, Red-headed Woodpecker, and Evening Grosbeak.

We peer at Horned Larks feeding on a grassy field. We earn three points for seeing the larks and four more points for finding a Lapland Longspur foraging among them. A Northern Harrier flies by on tilted wings, its white rump flashing as it goes. We earn two more points for seeing this splendid raptor. These birds are fun to see in their own right, but it is even more satisfying to earn points for seeing them, as it reinforces that what we are doing is an acquired skill that takes practice and study. We enjoy birding as a group, sharing in the experience of finding and identifying common and rare birds alike, and we look forward to sharing our sightings with the other teams at the end of the day.

The Mega Bowl of Birding in New Haven County is modeled after the Superbowl of Birding, held annually since 2004 by Mass Audubon as a fundraiser for the Joppa Flats Education Center in northeastern Massachusetts. That event attracts more than one hundred participants and generates thousands of dollars in donations to the education center. The organizers of the Superbowl of Birding have been supportive of our efforts to create a similar event here in Connecticut. The Mega Bowl is hosted by the New Haven Bird Club (NHBC) but is open to all.

Twenty-four people formed six teams for our inaugural Mega Bowl, held on Super Bowl Sunday in 2017. We observed a total of 103 species. The team representing the Connecticut Young Birders Club, the Darth Waders, earned the most points and had its name engraved on the Mega Bowl trophy. These enthusiastic young birders were thrilled to win the contest, outperforming several teams with more experienced birders. Their triumph lends hope to the future of conservation efforts in our community, as it is clear that the younger generation has its share of dedicated and conscientious individuals who appreciate nature. In 2018, twenty-eight people on seven teams observed over one hundred species, and again the Darth Waders achieved the highest score. The Mega Bowl has grown in popularity each year of its existence. In 2020, a total of forty-five birders on eleven different teams joined the fun.

Another team alerts us to a drake Harlequin Duck in West Haven. We follow that lead and earn five points for observing this beautiful bird. While there, we also collect four points for seeing an Iceland Gull and two points for seeing American Wigeon. We move on to the nearby Connecticut Audubon Coastal Center at Milford Point and spot a Snowy Owl roosting on a sandbar, far from its Arctic home.

The creation of a competitive activity is a natural consequence of any popular pursuit. People, through innate talent, interest, practice, and study, gain unique skills to accomplish certain tasks. Participating in a competition gives them a chance to demonstrate and improve upon their skills and compare them to the skills of others. Competitive events are created for virtually all known human activities, from intellectual pursuits to physical ones. Sports, games, and contests, from spelling bees to track meets, from quiz shows to figure skating, are all constructed to test people's skills in specific areas. Some events are ultra-competitive, with major prizes going to the winners and millions of dollars at stake, while others are for fun, intended to encourage participation and attract new entrants.

The Mega Bowl of Birding falls into the latter category and is one of many such competitive events available to birders. Perhaps the best known is the Big Year, in which birders strive to see as many species as they can in the United States in one year. Mark Obmascik wrote a book about this event, which was then made into a movie starring Steve Martin, Owen Wilson, and Jack Black. Similar activities include The Big Sit!® (created by NHBC member John Himmelman), and the World Series of Birding in New Jersey. Each offers birders the opportunity to practice their craft in a fun, semi-competitive, and rewarding way. The results of these efforts produce valuable information about the abundance or scarcity of bird species in various locations as well as changes in bird populations in specific areas over time. Some events also raise money that goes toward conservation and educational efforts in local communities and beyond.

By day's end, we've seen nearly seventy species and amassed 137 points. As darkness falls, we make our way to the Kellogg Environmental Center in Derby, where we will meet the other Mega Bowl participants. We are all tired, hungry, and cold, yet satisfied. We are a varied lot: high school and college students, doctors, teachers, environmentalists, retirees. Our ages range from six to seventy-nine years. We share a common purpose: to see and celebrate birds. Over dinner, we recount our rare sightings and dramatic moments. For the Winter Wrenegades, finding a pair of Snow Buntings on a rock by the shore was a highlight. For the Snowy Owlkins, a Bald Eagle flying overhead made an indelible memory. Each of us had a moment when we felt simultaneously in awe of and as one with nature. This is the essence of birding.

We hope that the Mega Bowl will be an annual birding event anticipated and enjoyed by a great number of birders for many years to come. The cumulative results will provide information on changes in the prevalence of bird species in winter in New Haven County over time. This will help us understand changes to our climate and natural environment. Recognizing these changes as they occur gives us the opportunity to mitigate their negative effects and enhance their positive ones, permitting us to preserve our natural spaces and the creatures that depend upon them.

Bird Walks and Those Who Lead Them

Ricci Cummings

From the inception of the New Haven Bird Club (NHBC) to the present day, members have enthusiastically led and participated in our Outdoor Programs. These programs are at the core of the club, and they provide members the opportunity to do some serious birdwatching together, guided by one of our skilled birders who is knowledgeable about the species found in a particular location. One such program is the bird walks; over the course of a year, the NHBC organizes approximately fifty walks throughout Connecticut and in neighboring states.

First Wednesday Walks

The First Wednesday Walks are especially popular, regularly attracting a crowd of members armed with their binoculars and scopes and their joy in birding. The Wednesday walks started in 2008, when a group of women who were avid birders and NHBC members were brainstorming new ways the club could engage its members. Betty Zuraw suggested walks, led by someone knowledgeable about birds and habitats, that would take place on the first Wednesday of each month, rather than on the weekends, as was our custom. Her thinking was that birding during the week might engage additional members in the club's Outdoor Programs and provide further birding opportunities for members already participating in the weekend walks. Penny Solum, Denise Jernigan, Sara Zagorski, and Kris Johnson thought the idea had potential, and Kris agreed to serve as coordinator. She subsequently identified suitable venues and lined up the leaders, and from that time to the present, the walks have grown in popularity and have expanded to a variety of venues.

Tina Green, who does not live in the New Haven area but for many years has been an active NHBC member, is currently responsible for organizing the First Wednesday Walks for our club. In that capacity, she selects leaders and locations for the walks, which typically last about four hours, although they can run longer. The walks are well attended, usually attracting at least twenty to twenty-five people, both novices and accomplished birders. All are welcome. Birders of all levels enjoy the useful exchange of pointers on sighting and identifying birds and the lively flow of information about the birds themselves.

We at the NHBC are blessed with an array of outstanding walk leaders who are widely respected in the Connecticut birding community. The elite leaders of this latest generation have mentored one another and are now mentoring the next generation on the First Wednesday Walks. Under their guidance, today's novices will become our future leaders, ensuring the sustainability of the NHBC and its mission.

Where We Walk

What makes an attractive venue for the walks? From a practical perspective, the walk coordinator chooses places that are accessible to the NHBC's members and have ample parking. But most important, she chooses a venue that is best at a particular time of year. For example, Sherwood Island State Park, Hammonasset Beach State Park, and East Rock Park are selected in May because of the profusion of warblers, such as Black-and-white, Prothonotary, Yellow-rumped, and Black-throated Green Warblers. In June, locations such as the White Memorial Conservation Center in Litchfield or Mohawk Mountain in nearby Cornwall/Goshen are great spots for viewing northern breeding birds. In September, the Coastal Center at Milford Point is a prime viewing spot for shorebird migration. Harkness Memorial State Park, situated on Long Island Sound in Waterford, Connecticut, has been a reliable choice in winter. A waterfowl hot spot, it features common species such as Mallards, American Black Ducks, Buffleheads, and Hooded and Red-breasted Mergansers, as well as infrequent ones such as Common Eiders and Black Scoters. Other popular walk locations include West Rock Ridge State Park, Lighthouse Point Park, and Edgewood Park in New Haven; Sleeping Giant State Park in Hamden; the lakes and the West River in Bethany; Bent of the River Audubon Center in Southbury; Naugatuck State Forest in Naugatuck—and that's just a small sample. The popularity of these walks turns in part on the rich diversity of habitats close to New Haven but also is grounded in the remarkable members who lead these walks. Generous with their time and expertise, they make our Outdoor Programs exceptional.

An Outstanding Walk Leader

Tina Green not only coordinates the First Wednesday Walks for the NHBC, she also leads the popular walks at Sherwood Island in Westport, Connecticut, about thirty miles southwest of New Haven. To learn more about what makes a successful leader for these programs, I spent a morning interviewing her. My goal was to learn more about how she got started birding; her various birding activities; and, of most interest to me, why she leads the walks and what she enjoys about them. All our many talented walk leaders bring their varying backgrounds, interests, and birding experiences to their programs. This is Tina's story.

What experience or insight first engages our walk leaders with birds? A personal encounter with a bird? An interesting article about birding? An association with a keen birder?

For Tina, it was a photograph of a Cedar Waxwing that she saw as a child. That photo captivated her and ignited an intense interest in birds. Tina is an artist with an artist's ability to note detail, shape, color, and texture, so it is unsurprising that for several years subsequent to that encounter, her interest in birds was focused on artwork and frequent visits to museums.

It was not until 2008 that Tina ventured out into the field and began her transition into a serious birder. She spent summers in Wellfleet on Cape Cod, and one day happened upon an article in the *Cape Cod Times* about a local bird walk sponsored by Mass Audubon. Intrigued, she thought she'd give it a try. The walk took place at Wellfleet Bay Wildlife Sanctuary, which is particularly rich in habitat, featuring woodlands, beach, heath, and salt marsh, and a multiplicity of birdlife. She had no binoculars, but that didn't deter her, and she was able to share the leader's scope. Tina recalls her astonishment as she gazed for the first time through that scope and experienced the thrill of an intimate, close-up view of a bird in its natural habitat. Profoundly moved and deeply engaged, she was, from that moment on, dedicated to observing birds and preserving their habitats. She purchased her first pair of binoculars and participated in all the walks scheduled for the remainder of that summer.

Back home in Westport in the autumn, Tina set about putting up bird feeders. She became a serious student of field guides and learned to identify birds ranging from Orchard Orioles to hawks to Ruby-throated Hummingbirds. With talent, hard work, and mounting experience, she grew from that novice into a leader.

Early in her birding career, Tina turned to the Connecticut Audubon Society and other groups to find organized bird walks. She walked with many accomplished birders, such as Luke Tiller, Penny Solum, Betty Zuraw, and Sara Zagorski. She had several mentors who were generous with their time and knowledge, including Frank Mantlik, A. J. Hand, Greg Hanisek, and Patrick Dugan. Tina credits them all with helping her hone her skills and encouraging her love of birding. In time, she felt confident enough to head out on her own, and she still loves the serenity of birding alone or with one or two people, which she describes as "meditative."

These days Tina leads about two or three bird walks a year. She volunteers in part because she is grateful for all the guidance and support she received while developing her birding skills and feels that this is an opportunity to do the same for others. Also, she simply enjoys sharing her love of birding with our members. She typically chooses to lead walks at Sherwood Island, which she reports is listed on eBird as one of the state's top spots for birding. In part because it features varied habitats, Sherwood Island can sometimes host up to sixty species in one day. Since she birds there four to five times a week, Tina is intimately familiar with the property and knows where and when to locate the various species. Common sightings include an incredible variety of waterfowl, gulls, and sparrows. Always well attended, the Sherwood Island walks allow for frequent and interesting sightings, and they enable Tina to pass her skills and knowledge down to others in the NHBC.

Like many of our elite walk leaders, Tina is well known in Connecticut and beyond for her dedication to birds and for her consummate birding skills. She is a member of several groups associated with birding and has served as both Vice President and President of the Connecticut Ornithological Association.

She is currently the only woman serving on the Avian Records Committee of Connecticut, which maintains the official bird list for the state by compiling and evaluating sightings for inclusion in the list. She not only compiles the sightings of others but also makes many herself: Tina found, for example, the first and third Western Meadowlarks in Connecticut.

Skilled Walk Leaders Find the Thrills

Birders puzzle over the good fortune of a rare bird sighting. Sometimes it just happens that way, and you're never sure why. Smart choices? Exceptional birding skills? Serendipity? Maybe a little of each? It doesn't really matter—when a rare sighting occurs, you just go with it.

Tina Green remembers one such occurrence. It was a cold November morning in 2015 on Sherwood Island, a diverse and productive bird habitat. Tina was leading a Saturday bird walk for members of the NHBC when she was able to relocate an Ash-throated Flycatcher that had been spotted earlier in the week by another birder. While making certain that everyone in the group was able to view the bird, normally found only in the western part of the country, she provided details about this flycatcher and explained the significance of the sighting. The birders who had gathered for the program were scarcely through celebrating their good fortune when another rarity, a majestic American White Pelican, made a fly-over! Even at a distance, it was breathtaking. Two unusual sightings on one walk was quite a thrill, and both the novice and experienced birders experienced an unforgettable day.

It's not just the rarities that capture interest. Tina discussed the pleasure she finds in all aspects of birding, as so many birders do. She gets the most enjoyment out of watching native birds that reside in the state year-round and are a constant part of our world, such as Common Raven, Killdeer, and Bald Eagle.

She describes her delight in seeing a raven on a nest again. She recalls the thrill of watching a bird do something that she hadn't seen before, recounting a recent sighting of eagles mating in an acrobatic flash of activity, in which the male climbed on the female's back for a brief but intense interval and then departed. She remembers the excitement she felt upon hearing a call she'd never heard before. She likes shorebirds because they tend to stay in one place and are easier to study. She loves the fall warblers, even though it's difficult to get a bead on them as they flit about. Our other walk leaders and members have their own favorites, both rare and common.

Why They Lead

What makes effective walk leaders, and why do they choose to lead? Judging from the experiences of Tina and other NHBC walk leaders, it starts with an encounter with the world of birds that resonates deeply and stokes an insatiable curiosity that propels the person to learn and experience more about birds around us. Sometimes a mentor nurtures that interest; sometimes it grows in solitude. It is typically someone with a good eye for detail and a good ear for birdsong. It is someone who will put in the time to learn, both from books and in the field. But expertise and aptitude alone are not sufficient. An effective walk leader is both a knowledgeable student who is constantly learning and a skilled teacher who finds pleasure in sharing that knowledge with others. It is someone who wants to give back. As exemplified by Tina Green and others who lead the club's walks, the heart of our walk leaders is found in their ability to experience sheer joy in everything associated with birds and the generosity of spirit to share that joy with their fellow birders. We in the NHBC are grateful for their contributions to our Outdoor Programs and hope our readers will join us on one of our walks to see for themselves why we are so proud of our walk leaders.

PART IV
Education and Conservation

Take Flight! Bird Observation with Children in a School Science Program and Beyond

Florence McBride

Black-capped Chickadee by Rob Fertiguena, a student in Hamden's Partners in Science Program. Florence McBride collection.

One day in early spring, a Hamden, Connecticut, third-grade class was watching Tufted Titmice, Black-capped Chickadees, and Yellow-rumped Warblers flitting around in the woods by the Mill River in East Rock Park. I was delighted that the students were seeing so many birds and excited because I had just heard my first Eastern Phoebe of the year. Suddenly, one of the boys exclaimed:

"This is the best day of my life!"

His words gave me one of the biggest thrills in my thirty years of work with teachers and students. The fact that the boy really meant what he'd said was shown five minutes later, when I suggested, "Someone said 'This is the best day of my life'—we should write that down," and he immediately took ownership: "*I said that.*" According to his teacher, the boy's parents would not have been likely to take him on a nature walk, so I was especially glad that he could have this experience with his class.

On another trip along the same path, a third-grade boy said, "Mrs. McBride, there's a Red-winged Blackbird," while I was talking briefly with a noted biologist. He had named the bird correctly even though it was not showing its red wing patch. After the biologist went on her way, I said to him, "Maurice, that was so cool!" and told him how special it had been for him to make the identification at that particular time. A few years later, near the first bend in that path, a fifth-grade girl observed, quite accurately, "Catbirds, catbirds, everywhere we go we see catbirds!"

These are just a few of the memorable remarks made during the field trips that have been part of my work as a specialist in the Hamden Public Schools Partners in Science Program. Site-based, with occasional field trips, the program is centered in outdoor observation of birds and makes many interdisciplinary connections. Working cooperatively with my partners, the teachers, I've had a wonderful opportunity to help students learn science skills through outdoor observation and to share my love of birds at the same time. Along the way, I've developed a large body of activities and materials, which are collected under the title *Take Flight!* and have been shared with many educators through workshops and correspondence. The program has received gratifying recognition from educators, scientists, and organizations, and there's detailed information about it in the Summer 2012 edition of the *Connecticut Journal of Science Education.*

Education and Conservation

A Hamden Public Schools Partners in Science field trip.
Florence McBride collection.

How did my chance to do this work come about? I'd been a birdwatcher in childhood, when the flutelike song of the Wood Thrush coming from the small patch of woods behind our house in suburban New Jersey made it my favorite bird for life. The first requirement for the Girl Scout bird badge was "Identify, out of doors, at least fifteen birds," and working on that was a lot of fun, especially when a spectacular male Rose-breasted Grosbeak turned up. I was lucky, too, that my parents fostered a love of birds and nature.

After this early start in birdwatching, my awareness and love of birds never disappeared, although other interests and responsibilities claimed most of my attention for many years. It was in 1980 that birding became a passion, when our family spent seven months based in Cambridge, England. Right outside our window there were beautiful species not found at home, and we saw many more during our excursions. I'll never forget the view from the hide at the Royal Society for the Protection of Birds' Minsmere reserve; it was like a waterbird diorama. Back in Connecticut, birding continued to be a big part of my life. I joined the New Haven Bird Club, spent a lot of time in the field, and took Noble Proctor's ornithology courses at Southern Connecticut State University.

Then in 1990, Paul Massey, the science director of the Hamden Public Schools, asked me if I'd like to work with four primary teachers and their classes in a newly created Partners in Science Program, to help them get some outdoor observation of birds into their programming. Paul was aware of my interest in birds because he'd gotten to know me during the latter part of my two decades of PTA (parent-teacher association) work in Hamden, when I'd interacted with many Hamden educators both informally and on numerous committees.

My first thought on hearing Paul's proposal was, "But I've never taught young students before—just college students—and I taught English, not science." But I'd relished my college science courses and really liked teaching, and I was quite familiar with the Hamden school system. So, I said yes, and immediately started to create two documents. One was a data sheet to use outdoors. The other was "Why Birds?" which shows how bird observation and related activities can help students learn science skills and concepts and develop an appreciation of nature, and how this focus can promote connections between school, home, and community.

It seemed a good omen that while scouting at West Woods School before my first meeting with a class, I found the only Connecticut Warbler I've ever seen! A class of Spring Glen six-year-olds spent its first time outdoors excitedly finding a multitude of things to observe, with Paul Massey in attendance. From then on, Paul's enthusiasm for our efforts continued to grow; he later wrote that Partners in Science was "the single most successful science program that I have been associated with in the over twenty years that I have been employed by the Hamden Board of Education."

That first year our program expanded from four of Hamden's elementary schools to six, and eventually it was implemented in all nine of them. It has never been a program that every teacher of a given grade has to participate in, but one that teachers can choose to join. As of 2020, after the retirements of many of its participating teachers, the introduction of No Child Left Behind, and an increased emphasis on standardized testing, it was still operating on a small scale and still bringing satisfaction to students and teachers, and to me.

The approaches, techniques, activities, and materials of the *Take Flight!* program can be useful for people who want to help children enjoy and learn about birds outside of school, as well as in formal educational settings. Here's some information about the program's basic assumptions and emphases, followed by a description of the way we conduct a typical bird observation class.

- It's important to realize that we don't have to go to parks or nature reserves, or be experts in bird identification, to start taking kids outdoors to look at birds. We can watch birds just about anywhere, and if we know how to observe birds and think about them, we can learn more along with the children. That can be a lot of fun, especially when a child can teach us something about a bird, or notices something we haven't, as when one Hamden first grader told her mother, "Quiet—I hear a chickadee!"

- Since observation is the key, children with many learning styles and different levels of other skills can be successful bird observers. Teachers have commented on the way our activities can promote self-esteem. Some of the most significant beneficiaries of the Hamden program have been students with special needs, many of whom have made very perceptive observations. We've found that some students show signs of becoming serious birders, like the first grader who identified a Cattle Egret in a large mammal's pen at the Beardsley Zoo in Bridgeport. He must have been studying his field guide!

- Children can communicate their observations in many ways: spoken comments, writing, drawing, and modeling. These often show admirable observation abilities and express enthusiasm for birds.

Gull by Mark Mancini, a student in Hamden's Partners in Science Program. Florence McBride collection.

- The program has interdisciplinary components, and the children's enthusiasm can promote learning in many subject areas—reading and language arts, mathematics, art, music, and social studies—as well as motivating students to do research.

- Practical matters: We want to help children learn some basic outdoor health and safety practices, like avoiding poison ivy, being careful not to look at the sun (especially when watching flying birds), not approaching wild animals or touching mushrooms, and checking for ticks after a bird walk. We set guidelines for appropriate behavior during our outdoor explorations, and stress that we want to observe nature without changing it, not picking plants or chasing animals (including insects).

- Although helping students learn science skills and concepts is our main rationale, there's another very important goal of our activities. We want to help children develop an appreciation of the fascinating and beautiful natural world, and a sense of its importance in our lives.

Here's how our basic bird observation lesson works.
We always do indoor preparation before taking our classes outdoors.

We typically allow more time for this during a class's first session than during later ones. We find that even a brief preview can enhance a walk with young "bird detectives." Bird photos from calendars, posters, field guides, apps like Merlin (free from the Cornell Lab of Ornithology), websites like Cornell's All About Birds, and other references, including videos and audio tools, can help with this.

Indoors, away from outdoor distractions, we introduce the kids to learning about:

What birds look like: Showing children pictures, our life-size two-dimensional *Take Flight!* bird models, and/or videos of a few of our most common birds can lead to thrills of recognition outdoors. The visual teaching aids can also serve as references for coloring bird pictures, which are widely available in publications such as the Dover coloring books. In addition to looking at color and pattern, it's a good idea for children to start to learn about the sizes and shapes of birds. To get a feeling for size estimation, they can look at life-size silhouettes made for the program, comparing them with our models and holding their hands apart to get a sense of the length of large (crow), medium (robin or jay), and small (sparrow) birds.

Bird sounds: Our program stresses these; so often they're our only clue to a bird's presence, as well as being an important aspect of bird behavior. Knowing some calls and songs is an expertise that children can develop and enjoy. Young students like to imitate some of the sounds—a good way to start learning them. My DVD *Some Bird Sounds of the Northeast* helps with this auditory learning; beginning with the crow, jay, and chickadee (a great one to imitate), it shows the birds making the sounds.

Other bird behavior: We don't usually talk a lot about this before going outdoors, although my video introduces some bird behavior, as does my "Blue Jay Rap." Students have loved to have me perform the rap, which conveys some facts about Blue Jays as well as helping to create enthusiasm. The kids participate by snapping their fingers to set and keep the rhythm, and they can learn and perform the rap themselves later. There's a "Chickadee Rap" too. (Soon after our introductory lesson, we often do one called "What are the birds DOING today?" which focuses on bird behavior. For this activity, we typically record a large number of predictions indoors, and then observe outdoors, using data sheets designed for the lesson.)

Some basics of successful bird observation: To promote attention and avoid scaring the birds, we establish a few procedures, often by asking the kids questions so that their answers can help set the rules. We tell them we should stay together, so that everyone will have the best possible chance to see the birds we find, and no one should get in front of the leader. We should talk quietly (partly because we want to hear the birds!) and not make sudden movements or run around.

Because it's important to learn to recognize poison ivy, students are shown a picture of it before we go outside. Coloring pictures can help us learn what things look like, so all our students have my poison ivy coloring picture in their interactive bird books, which were created for the Hamden program.

Scientific observation: Since ours is a science program, we introduce age-appropriate procedures for this. We typically make a simple prediction (hypothesis) about what we'll find outdoors and check it against our observations when we get back inside. Students are instructed to ask me and each other, "How do you know that?" when they make statements about the identification of the birds we find or interpret their behavior. Teachers or students record our observations, including time and weather data as well as information about the birds found.

After this preparation, we go outdoors!

As birders know, morning is generally the best time to look for the land birds we're most likely to see during the school day. We usually try to get outside as early as we can.

We have a technique for helping groups of students get as close as possible to birds when we're in an open area like a schoolyard or playing field; this is especially important if we're not using binoculars. So that we will all move and stop at once, the leader establishes a four-beat rhythm, and we then take three steps toward the bird and "*Freeze*" on the fourth beat to observe. After being sure the bird is relaxed, we repeat this process. To keep the bird from becoming alarmed, we don't walk straight at it, but tack toward it in a zigzag pattern. It's amazing how close you can get to many birds this way.

As we watch and listen to birds, we can think about questions like:
- Where is each bird, and where within its habitat? For example, woodpeckers and nuthatches are the birds most commonly seen on tree trunks, and they're typically in different positions: woodpeckers head-up, nuthatches

often head-down, as shown with this *Take Flight!* model placed outdoors for students to observe.

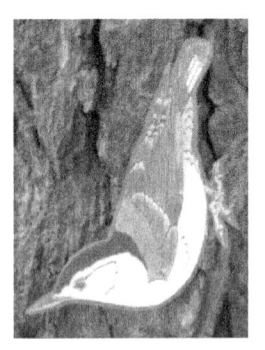

Nuthatch model created for the *Take Flight!* program.
Florence McBride collection.

- How big is each bird, and what do we notice about its shape?
- What are its colors, and what patterns do the colors make?
- How does it move (e.g., walking vs. hopping, flight style)? Young children often enjoy imitating the way birds move.
- What are the birds doing? How are they fulfilling their basic needs?
- How are the birds we're observing alike and different? Especially as children become more experienced observers, we want to introduce them to scientific classification in an age-appropriate way, with the concept of bird families.
- What other interesting things can we notice or find?

We take time to appreciate what we're observing. When children say things like "Ooh—it's beautiful!" (he was looking at a cardinal) or "This is the best thing we've done!" it seems clear that the study of science and nature is giving them an opportunity to enrich their lives in a way that goes far beyond the purely academic value of what we're doing.

I'm sometimes asked: What about binoculars? Although binoculars unquestionably enhance bird observation for many children, they are not necessary for *Take Flight!* activities. In our program, we don't use them until we've conducted our binoculars lesson, which we do in spring with second graders and older students. We've found that most of our students can use binoculars and like to do so.

Bird observation can enhance and energize education because of the way it can engage and motivate students. For families it can be a wonderful way to spend time together. Many of us birders would agree with the second-grade boy who said, during one of our outdoor explorations, "I wish we did this every day!" and with this exclamation from a first-grade girl (who during an earlier session had complained cheerfully, "My head hurts with all the thinking"):

"I was born to be a birdwatcher!"

Blue Jay Rap

He's a **bad Blue Jay**, and I'm **here to say**
You can **see** this **bird** almost every **day**.
When **winter comes**, he'll **be here still**,
Hey, look! He's got an acorn, **in his bill**.
In **spring** time **he** may **rob a nest**,
But **remember, sometimes he's the best!**
'Cause **when** he sees an **owl** or a **cat**
He **sounds** as though he's **yelling, "Scat! Scat! Scat!"**
He **tells** the other **birds** when **danger's near**,
And his **warning's** so **LOUD**, that it's easy to **hear**.
He **helps** them **get away in time**,
And he **doesn't / even / charge a dime!**

Text © 1990 Florence S. McBride.
Drawing by Mike D'Ascanio used by permission.

"Blue Jay Rap." Florence McBride collection.

Restoring Local Populations of American Kestrels, One Box at a Time

Tom Sayers

Beginning with my earliest wanderings as a youth, I have had a special admiration for the American Kestrel. Over the past thirty years, the number of these little falcons has been very steadily declining in Connecticut. There are several factors undoubtedly contributing to the decline in kestrel numbers throughout the Northeast, but the truth is that no one knows for sure exactly how significant any one of these factors is.

It is posited that some of the factors adversely affecting Northeastern kestrel populations are, in no particular order, pesticide use, loss of suitable nesting cavities through the removal of dead and dying trees, competition from starlings for suitable nesting cavities, loss of suitable grassland habitat to development, and predation from the increasing numbers of Cooper's Hawks.

The author begins the March ritual of cleaning out American Kestrel nest boxes and adding new shavings. Photo Northeast Connecticut Kestrel Project.

Before I began working with kestrels, I was very actively involved in putting up nest boxes for bluebirds. As I perused the literature on bluebird nesting boxes, I would occasionally come across articles describing nest boxes that were produced specifically for the American Kestrel. I was quite skeptical at first that a member of the falcon family would use a man-made nesting box, but as time went on, I became more and more intrigued by the idea. At about that time, I had the good fortune to meet Art Gingert, from West Cornwall, Connecticut. Art has had a successful kestrel nest-box project in Connecticut for over thirty years. After hearing firsthand from him about his success, I decided, in 2009, to try to actively work as a citizen scientist to begin rebuilding local populations of the American Kestrel by starting my own nest-box program. Art's mentoring early on was crucial to my eventual success.

My study area is approximately 120 square miles, encompassing the north-central Connecticut towns of Somers, East Windsor, Ellington, Enfield, South Windsor, and Mansfield. In the beginning years of the project, I encountered a very steep learning curve. The subtlety and the number of variables that go into successful nest-box placement, and the most effective methods to mitigate the competition from starlings, seemed initially overwhelming to me. I made many, many bad decisions on box placements in the early years. My records indicate that, since my project's inception, I have taken down more than 140 nest boxes that were in poor locations. Over the years, I began paying much closer attention to the very specific habitat requirements that were necessary for kestrels to experience nesting success. This resulted in a lot of trial and error, but slowly my efforts began to pay off. Even in my first year, two boxes, out of a total of seven, successfully produced seven nestlings that fledged. I was hooked. During the 2017 breeding season, forty-five out of eighty total nest boxes produced 173 nestlings.

A male American Kestrel who took over the care of the young when the female was injured, photographed in Storrs, Connecticut.
Photo Northeast Connecticut Kestrel Project.

What became obvious to me over time was that I wanted my project to be more than simply an effort to produce more kestrels in my study area. I realized that when studying any species in serious decline, accurate, reliable data is crucial to helping us understand the underlying forces at work. I became committed to better understanding the population dynamics of the kestrels in my study area. To that end, I began collaborating with the Connecticut Department of Energy and Environmental

Protection (DEEP), the University of Connecticut, and the American Kestrel Partnership on many different population studies focusing on the kestrels in my area.

Beginning with the very first nestlings back in 2009, I have banded all nestlings with federally approved, metal leg bands. Each of these bands is engraved with its own unique number, thereby identifying that particular kestrel for life. Beginning in 2012, I also began a concerted effort to trap and band as many adult birds as possible. For these federal leg bands to provide any meaningful data on the birds' dispersal patterns, survival rates, and degree of site fidelity, the marked birds need to be recovered so that the band numbers can be read. There are several ways banded kestrels can be recovered: 1) someone finds a dead bird and reports the band number to the federal Bird Banding Laboratory; 2) the birds occasionally are trapped and then identified at various raptor-banding stations located throughout the Northeast; or 3) the birds are trapped by me, in either a bal-chatri (live-prey-baited trap with filaments to entangle raptor legs) or a box trap.

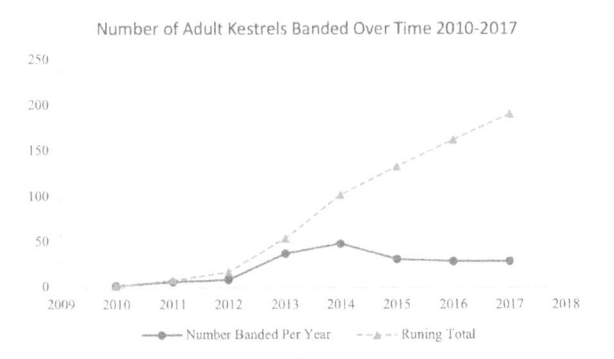

Number of adult American Kestrels banded over time.

Average Brood Size 2011-2017

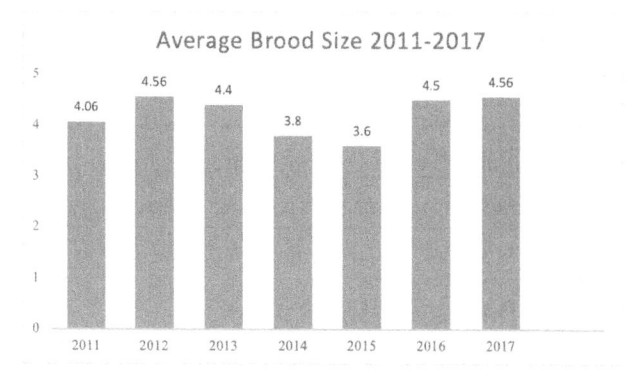

Year	Average
2011	4.06
2012	4.56
2013	4.4
2014	3.8
2015	3.6
2016	4.5
2017	4.56

Average American Kestrel brood size over time.

In addition to federal leg banding, I also spent two seasons carrying out a radio-telemetry project involving only young birds that were about to fledge. These nestlings were outfitted with very small VHF radio transmitters that emitted a radio signal unique to each bird. To track the birds, a volunteer had to cover very large areas with a handheld antenna that would (hopefully) detect the radio signals and allow the birds' positions to be plotted over time. The intent was to focus on the patterns and timing of young birds' dispersal from their natal area. This would, we hoped, provide us with information to better prioritize land-management practices for areas determined to be of highest use during post-fledging dispersal.

During the 2014 breeding season, I received a grant that allowed me to collaborate with the Connecticut DEEP on a two-year project involving the deployment of leg-mounted geolocators on adult birds. Unlike the VHF radio transmitters I referred to earlier, geolocators do not transmit signals but instead store thousands of ambient-light data points that can be triangulated to obtain the latitude and longitude of each bird at discrete points over time. In contrast to relatively short-term data gathered from VHF radio transmissions, these geolocators can provide data points covering the birds' movements for a period of over a year. The confounding factor in obtaining the data, however, is that, just as with federal leg bands, the birds need to be recovered after the unit's deployment in order to download the data.

Below I have attempted to highlight, in general terms, some of the key findings that resulted from the various field research projects described above.

Federal Leg-Band Findings

This is only a partial listing of the extensive findings that have resulted from analyzing federal leg-band data over the past eight years.

- Adult American Kestrels will routinely return to within five miles of the box they successfully used the previous year but will very rarely use the same nesting box twice over time. I have been able to document only two instances of adult kestrels using the same nest box more than once.
- Over 95 percent of all recaptured kestrels in my study area were originally banded by me in the study area. Foreign recaptures—birds that were originally banded outside of my study area by someone else—are quite rare.
- Birds from my study area have been recaptured as far away as Florida.
- Approximately 60 percent of recaptured birds in any year were originally banded as adults in the study area.
- Approximately 40 percent of recaptured birds in any year were originally banded as nestlings in the study area.
- Newly banded nestlings in my area have been recaptured as far away as Cape Cod only six weeks after fledging from their box.

- In any given year, recaptured birds constitute approximately 60 percent of all adult birds captured. This indicates that approximately 40 percent of all adult birds in any given year are unbanded adults and probably new to the area.

Radio-Telemetry Findings

- After nestlings initially fledge from the nest box, they seek the shelter of nearby trees and will remain hidden in the canopy for ten days or more. This period is characterized by very short forays from the canopy and overall limited movement. The fledglings rely almost exclusively on the adults for feeding at this time.
- From two to four weeks after fledging, the nestlings begin hunting on their own and making longer flights of up to three hundred yards away from their natal site. This is when they can become much more vulnerable to predation and conflict with man-made objects—including airplanes. American Kestrels are the number-one bird species involved in plane strikes in the United States.
- Sometime around the end of August or early September, the young birds, in loose association with their siblings and parents, begin migrating south. At this point they can make unpredictable one- to fifteen-mile (or longer) flights completely out of their home territories. The limits of radio-telemetry technology make it very difficult to accurately track them as they embark on these longer journeys.

Geolocator Findings

- Since geolocators enable us to re-create the birds' movements for up to a year, we were able to determine that three of the four kestrels we followed ended up wintering in Florida south of Miami. One of the birds stopped earlier, wintering in northern Georgia. On their return trip north the following spring, they all followed a remarkably similar path, in reverse, to each of their individual routes south in the fall. Since these birds were recaptured so that we could retrieve this information, we know that they made a twenty-seven-hundred-mile round-trip and ended up back within a mile or two of their previous summer's location in the study area.

- The kestrels studied all took routes while migrating that generally remained within one hundred miles of the coast.

- Geolocators also record conductivity (a measure of the moisture level present in the bird's environment) several times a day. It was very interesting to document frequent periods of high conductivity throughout migration. This indicated an affinity for high-moisture areas in their environment, at least during migration, and suggests that bathing of some type could be a regular occurrence. This finding has previously been rarely documented.

- Abrupt decreases in light levels usually occurring near the end of the day suggest that, even in migration, kestrels often rely on man-made structures for nighttime roosting.

So, there you have it. I have tried to present a fairly comprehensive overview of my work with kestrels. This work is never about me. I see it as an example of what any motivated individual can accomplish when committed to making a difference with any species in peril. One person, and that next person could be you, *can* make a difference. Due to the combined efforts of Art and myself, the status of American Kestrels in Connecticut has now been upgraded from threatened to species of special concern. It doesn't get any better than that.

A Condo Development That Bluebirds Love

Pat Leahy

In the mid-1980s, I discovered birdwatching. I had started my working career as a special-education teacher and had been a great fan of nature and lessons that got us outdoors. My initial birding experiences were a few Connecticut Audubon walks, and then I happened upon the New Haven Bird Club (NHBC). The president was Carl Siebecker, one of the friendliest fellows you would ever want to meet. It did not take long before Carl had me participating—and very active—on the club's Board. At the time, Milford Point was transitioning from NHBC stewardship and developing into the Connecticut Audubon Coastal Center at Milford Point that birders enjoy today. I was the club's liaison to the Coastal Center's advisory board, taking over the position from Fred Sibley. As Fred had done for many years, I focused my stewardship at Milford Point on trying to keep the invasive plant species from overtaking this gem of a property. The wonderful payment was an intimate experience watching the seasonal changes and birdlife. I was given the formal title on the advisory board of "Chairman of Building and Grounds." That meant I had to keep the lawns mowed!

A few years into my adventure as a birder, my wife, Cathy, and I decided to move from Milford up to Bethany. It got a little harder to spend so much time at Milford Point, and with the Coastal Center up and running, I decided to find some opportunities in our new environment. When the Regional Water Authority (RWA) implemented a pass system for hiking at Lake Bethany and Lake Chamberlain, I was quick to join up. Lake Chamberlain in spring is a mecca for Tree Swallows, Rough-winged Swallows, Barn Swallows, and my favorite, the Eastern Bluebird. There had been an earlier effort to put some Eastern Bluebird/Tree Swallow nest boxes at Lake Chamberlain, but the boxes were not being maintained. Through the NHBC I had gotten to know John Triana, the club historian and a past president, who worked for the RWA. I let him know about my skills as a woodworker and my desire to learn how to develop an Eastern Bluebird/Tree Swallow trail. As a test run, in my first year I was given permission to erect boxes on some farm fields maintained on Downs Road in Bethany.

Eastern Bluebird, resident of Lake Chamberlain in Bethany, Connecticut, sitting on a Regional Water Authority testing pipe.
Photo Bill Batsford.

Education and Conservation

Each year I would ask permission to add another of the RWA properties to my trail. The area below the dam on Lake Dawson looked to me to be just perfect. This area is not open to the public, but it can be viewed from the Woodbridge Land Trust's Bishop Estate East Trails that surround the Thomas Darling House in Woodbridge. I remember heading over in late March to establish a row of boxes. When I tried to drive the posts into the ground, I would get six inches in and hit a solid stop. I assumed the ground was still frozen. I waited until late April and tried again. What I thought was ice turned out to be solid ledge. The RWA had scraped the area when building the dam and put only six inches of topsoil back. To place my boxes on the property, I had to find cracks in the bedrock. I figure it took about sixty tries to put six boxes in. With very sore arms, I decided the six boxes would work just fine.

I also was given permission to take over the abandoned boxes at Lake Chamberlain—an amazing place. When insects hatch from the water in the spring and summer, they provide an absolute smorgasbord of food for the swallows. I discovered I could put lots of boxes close to one another, and there would still be plenty of food for all. This setup also presents the public with a great opportunity to see these amazing creatures "up close and personal." I have led walks on late spring days when the chicks are just fledging. It doesn't matter whether you are four or ninety, you will get excited and smile as you watch the antics of the new family.

I have learned that it takes a brood about three weeks from egg laying to hatch and three to four weeks to fledge. I am vigilant about cleaning the boxes in preparation for the new season, but I try not to bother the birds during the nesting season. I have also learned that there is an unusual real estate market in the bird world. If I put up ten boxes, at most I get one or two

Eastern Bluebirds and eight or nine Tree Swallows. Occasionally, a House Wren competes for space. I have frequently found a House Wren's nest on top of a Tree Swallow or Eastern Bluebird nest from earlier in the season. The House Wrens are a bit of a nuisance, because the male builds several of these stick nests and then takes his mate around to pick a suitable home. The boxes with the rejected nests are then unusable by anyone until we get them cleaned out.

With a reputation for being the "Bluebird Guy" at Chamberlain, I was asked, along with NHBC member Chris Loscalzo, to join a Bethany Conservation Committee meeting on the farm owned by Peter and Diana Cooper that abuts the lake. More than forty years ago, the Coopers and friends of theirs, Guido and Anne Calabrese, purchased a 110-acre farm that overlooks the West River valley and split the land into two farms. These two families did not want to see this pristine farmland turned into building lots. Peter Cooper was applying to have a significant portion of the farm deeded to prevent development. Documenting some significant species living on the land would be helpful to the process. It was an opportunity for Chris and me to bird a beautiful and productive farm while showing off the wildlife to the conservation committee. The payout was a lovely day in the country that would help save a treasured plot of land.

My favorite moment of the day happened as we approached a wooded area near the fence along Lake Chamberlain. I commented that this was a great place for all kinds of woodpeckers. As if it had been rehearsed, a Pileated Woodpecker flew in and landed on a stump and started drilling for grubs. Luck had sealed my reputation with the group! Since then, the Coopers and Calabreses have generously allowed us to run a number of NHBC walks on the property. We have also included the property on our summer breeding-bird count and our Christmas

Education and Conservation

Bird Count. From the top of Peter's apple orchard, one has a quite amazing view of the shoreline of Long Island!

I have never been a driven birder, trying to build that three-hundred-plus life list. I have preferred to head in two different directions. I really enjoy studying the whys and patterns of bird behavior. Although my surveys are not totally scientific, I have watched the numbers of bluebirds and swallows in our NHBC counts stay high or even grow in a time when we are hearing about their decline.

I also absolutely enjoy watching people get excited at seeing the beauty of our world. I have led a couple of walks a year for at least the past ten to fifteen years, and the RWA lets me take people on sites that are normally closed to the public. Watching people as they observe an Eastern Bluebird family, Spotted Sandpipers along the edges of the reservoir, or a bunch of crows mobbing a young Bald Eagle is a hoot.

I believe I have in excess of seventy Eastern Bluebird/Tree Swallow nest boxes now. When an ornithology professor asked me what percentages of my houses were being used, my response was "all of them!" Mike Ferrari and DeWitt Allen have been my faithful partners in keeping the nesting boxes up to snuff. I remember putting up boxes with DeWitt on Lake Watrous. As we finished one box, we would look back and see that Tree Swallows were already inspecting the previous one for occupancy. The RWA's desire to encourage diverse environments on the land they steward, along with the NHBC's mission to conserve natural resources, and my willingness to spend some pocket money on some wood, screws, and fence posts have added up to one of the great joys in my life.

PART V
Tributes

There's Always Something to See: A Tribute to Noble Proctor

Dan Barvir, Frank Gallo, Patrick Lynch, and Florence McBride

Noble S. Proctor was a university professor, wildlife biologist, distinguished author, and a world-traveling naturalist and tour leader. He was also the best, most supportive—and funniest—teacher most of his students had ever known. Over almost half a century of professional activity, Noble was beloved by his students and friends in Connecticut, by his birding buddies from the New Haven Bird Club (NHBC) and other groups, and by many acquaintances throughout the world. In his thirty-four years of teaching ornithology, botany, mycology, and biogeography at Southern Connecticut State University (SCSU), Noble educated generations of students, instilling not just facts and figures but a lifelong love of the natural world that they have carried far and wide in their careers and personal lives. Noble wrote or coauthored ten books and was an accomplished wildlife photographer. In 2013, the American Birding Association gave Noble its Roger Tory Peterson Award, the most prestigious honor in American birding. As this description illustrates and as those who knew him would agree, Noble was aptly named. A birder of the highest class, he acted on behalf of birds and the natural environment, demonstrating to his fellow birders, students, and friends what it means to make a difference in others' lives.

Working with his close friend Roger Tory Peterson on multiple editions of Peterson's *Field Guide to the Birds of Eastern and Central North America*, Noble helped research and clarify the species range maps for most North American birds. Peterson once called Noble the best naturalist he had ever known. Noble was later one of the founding members establishing the Roger Tory Peterson Institute for Natural History in Jamestown, New York. "You could drop him anywhere on the planet, and he could tell you what [bird] you were looking at," said Twan Leenders, the institute's director. "Everybody was in awe of him. He was always the guy in the back of the bus. There was no pretense about him. He would always be encouraging and inviting and helping you figure things out."

Noble Proctor. Photo Dan Cinotti.

Noble Proctor grew up in Ansonia, Connecticut, where he described roaming widely in the Naugatuck and Housatonic valley farmlands and parks, fascinated not just by birds but by every aspect of the natural world, from ferns and fungi to insects and other animals. Reminiscing years later, Proctor recalled wandering in local fields and forests as a grade-school student, long before he owned binoculars or even had many reference guides. "I looked at everything—everything was interesting. If there weren't birds around, I'd do butterflies, bugs, or wildflowers. I knew them through experience long before I knew their names." A local greengrocer was the source of more exotic wildlife: "This was before the days when they would spray everything coming in. They'd get these giant bunches of bananas, and you could never tell what sort of thing would crawl out of them. That's how I saw my first tarantula and a lot of other Central American stuff. I'd help out in the store, and they'd let me keep whatever bugs crawled out of the fruit."

Proctor attended Ansonia High School and, upon graduation, entered the United States Army. After his army years and before starting college, Noble was employed by Yale University to collect materials for protein and genetics studies for some of the first DNA-based taxonomy studies, precursors to those that have since revolutionized ornithology and our view of the evolution of birds. He received his bachelor's and master's degrees at SCSU and his PhD at the University of Connecticut. Because he was already a skilled birder with a national reputation when he started graduate school, even many close friends were surprised by Noble's decision to do his PhD thesis in phycology, studying Connecticut algae, but Noble always maintained that he was at least as interested in plants and fungi as he was in birds. Longtime friend and field botanist Margaret Ardwin said of Noble, "He was an old-fashioned naturalist. Today, people specialize more. He was enthusiastic about everything."

Noble's teaching influence extended far beyond his own SCSU classroom. Wildlife educator Florence McBride says,

> It's impossible to overstate how important Noble's help was to me as I developed and implemented my *Take Flight!* program in the Hamden Public Schools. Without his inspiration and his teaching, and his ongoing encouragement and support, my efforts to help children and adults learn about and appreciate birds would never have gotten off the ground. When I began working with Hamden teachers and their classes to get some direct observation of birds into their programming, I asked Noble to take a look at the materials I was developing. His enthusiastic response gave me the courage of my convictions. For the rest of his life, he was always available for consultation, and at various times he affirmed his support in other ways, first by getting me invited to a workshop at Cornell University, and later by asking me to help with a weeklong summer workshop for teachers and using my bird models in a workshop that he gave.

McBride also accompanied Noble on international tours to Africa and Central America. "Noble was kind to everyone. As a friend once said to me, you never heard him say anything negative about anybody. On the Kenya and Costa Rica trips I had the good fortune to experience with him, he knew what each participant was interested in and helped us implement our priorities, waiting patiently to see a species that was calling from a dense forest, making opportunities for netting insects, and being sure that we were well positioned to photograph or film the wildlife we were watching."

Skilled naturalist and regional birding expert Frank Gallo, a former student of Noble's at SCSU, remembers a favorite tour story from Africa. "Noble's tour group was in a Maasai village, and Noble thought he would impress them by walking up to one of the warriors and talking with him in Maa, the local language. Noble left the group and walked a distance up to a red-sashed warrior standing outside a hut with his spear and said "hello" in Maa. The guy looked at Noble and said, in perfect English, "So, how are the Yankees doing?" In the retelling of the story, Noble said he almost fell over in surprise!

It turns out that the former Maasai villager attended Yale. Noble asked, "What are you doing here?" The gentleman responded, "Oh, my dad likes it when I come by for a visit." They chatted for a while before Noble proudly sauntered back to his group, whose members were out of earshot and, of course, now stood in awe of Noble's Maa language skills. I don't think he told them the full truth about the encounter until later, when he shared the joke with the group."

But beyond Noble's celebrated stories and travels, Gallo most appreciated Noble's positive outlook on life. "There was something about Noble, a joie de vivre. It wasn't just his kindness or impish grin, his dry sense of humor, or his marvelous stories. It was his spark. Noble was comfortable with who he was, and you could see it in his twinkling eyes."

Longtime East Rock Park ranger Dan Barvir was another student strongly influenced by his experiences in Noble's SCSU classroom.

> From day one, I was impressed by this man's seemingly boundless energy and wealth of knowledge and knew I would be taking more classes with him. Though the classes were challenging, the way Nobel taught by telling of his travels across this planet was both entertaining and enriching. His Bio-Geography class was one of my favorites. Who needs Sir David Attenborough to tell about the Earth and all its life forms, when we had Noble Proctor? Though I have never been on one of his birding tours, we heard him speak at New Haven Bird Club meetings; his words always made you feel like you were there.

Although always encouraging and supportive, Noble was no pushover in the classroom. Author and artist Patrick Lynch remembers Noble's classes as fascinating but very demanding: "Noble made you want to do well, to bring your knowledge of a subject up to his high standards. It wasn't about getting a perfect test score; it was always about truly sharing Noble's enthusiasm for the natural world."

Pat Lynch recalls Noble's deep personal engagement with students and his remarkable memory.

> I started at Southern as an art major, and after an unhappy semester, I decided to switch to biology. For guidance in the new major, I was assigned to a young faculty member, a Doctor Proctor. We met, and Dr. Proctor turned out to be a helpful and patient guide to starting over as a biology major, and after a pleasant half hour or so I went on my way. About six months later, I was walking between classes in Morrill Hall, not paying much attention to what was around me, and I suddenly heard, "Hi, Pat, how's it going?" as Noble passed me in the hall. Of course, I remembered Noble, but I was stunned that among all the hundreds of students with whom he interacted, Noble had recalled me six months after a brief conversation.

But that was Noble. Every one of his students has at least one of those stories; some have dozens. "When you're in your late teens or early twenties, you don't necessarily have the language to communicate what's so memorable about the few people that you meet who are truly extraordinary, but even when you're young and inexperienced, you know it when you see it. 'Centered' is the word I would choose today," Lynch reflected. "In a chaotic world, few people have such a strong and certain sense of self—of their mission and place in the world. Those who do stand out like lighthouses in your life's journey. That was Noble."

Noble gave his students access to the beauties of the natural world, but decades later what his students remember most are the profound life lessons bound up in Noble's approach to the living world, and his humane, holistic view of our place in nature. Deep engagement, patience, and a keen awareness of the world around him constituted Noble's approach to birding and to life. Noble once said, "There is always something to see. That's the great part of it. You are not going to see anything unless you get out there and take a look. Persistence is the key: Don't give up. Don't do it fast and walk away from it. Go in, spend time, and look." Noble lived that philosophy, and decades later, his many students recollect not just his remarkable breadth and depth of knowledge of the natural world but also Noble's humor, grace, and puckish sense for the ironic.

Frank Gallo remembers:

> I was walking up to the Branford hospice in the afternoon on the day Noble died [in 2015], when I noticed that all the upper windows of the hospice were whitewashed with bird droppings. Sitting on the railings and all over the roof of the building were pairs of Herring and Great Black-backed Gulls. They were nesting on the hospice roof. I started to laugh, and I thought to myself, "Well look at that, Noble, you old fox, you're going to die in a gull colony." How perfect. You really can't make this stuff up.

The Gifts of Richard English

Martha Lee Asarisi and Michael Horn

At a glance, you might not understand what was so special about Richard "Dick" English. If you met him—dressed casually in a plaid shirt and baseball cap, carrying a little black lunch box or a pair of binoculars, with his typically humble manner—you might conclude that you were encountering a run-of-the-mill tourist, without any particular birding interest, visiting New Haven. And would you ever be wrong! Through his lifelong interest in birding, the impact of Dick English in the local community and birding circles has been truly wide and deep.

A Chance Encounter Leads to Birding

Martha Lee Asarisi remembers her chance encounter with Dick English in 1988, when she found herself sitting next to him at a summer solstice picnic in Milford, Connecticut.

Newly married to Richard Asarisi, she was also new to her home in Hamden, Connecticut, and had just accepted a new position as staff pharmacist at the Yale University Health Plan. Her reaction to all this change can be summed up in this statement made to a New Haven Bird Club (NHBC) member: "Everything was so new to me that [it] was making me more than a little anxious."

As her husband was chatting with others at the picnic, she found herself silently enjoying a spectacular view of Long Island Sound—albeit still feeling somewhat uncomfortable. Martha continued her story: "A nice man—Richard English—sitting next to me started chatting about how sunshine on a summer solstice was rather fitting."

With the ice broken, her anxiety lessened. They conversed about his early childhood home on Hillhouse Avenue in New Haven, his current home near East Rock Park, and her new job and responsibilities. The time passed. Much to her surprise, she learned that the Yale Health Plan parking lot was located in his former backyard!

Then the conversation turned to the topic of birding. Dick English asked her if she liked birds. When she replied that she didn't know much about them, "that lit a spark in his eyes like a child's on Christmas morning." Before she could interrupt to say "birding is not for me," they—along with nearly every guest at the picnic—were strolling their way down the road to the Milford Point Audubon property for some exploration.

"Richard and I were leading the way with gusto and enthusiasm!" she exclaimed, remembering this unanticipated occurrence. During the walk, Dick explained that the fragility of Long Island Sound bird habitat was caused mainly by human intrusion into important breeding areas. He was concerned, and his fear for the birds was genuine and contagious.

While at Milford Point, Dick English checked around for the nest of a bird that, Martha recalled, he said "would make me a birder for life because of how special it was." Without binoculars and guided only by his lifelong knowledge of birds and habitat, he found the nest on the beach. Hidden in plain sight, what appeared to be just a collection of gravel on the sand proved to be camouflaged eggs.

"Careful, the bird must be somewhere close," he whispered.

Martha continued: "We all stopped, and he asked me to squat down and focus my eyes on orange legs, and all of a sudden the sand came to life. I saw my very first Piping Plover!"

Ever since then, Martha annually visits Milford Point to see the Piping Plovers (now categorized as threatened in Connecticut). Her perspective about meeting Dick English was expressed in the following: "I gained this hobby I really enjoy, all because of a chance encounter with one of the greatest birders I've ever had the pleasure of meeting."

Thanks to Dick's willingness to engage with someone unfamiliar with birding, her passion for birding commenced that day. Thirty years later, Martha is a deeply committed birder. She is active in both the NHBC and the Connecticut Audubon Society, and she serves as an Osprey steward for the North Haven, Connecticut, area.

"Richard English was one of those guys that—even if you never met him before—his zest for life and passion for nature shone so brightly, that at first meeting you couldn't help connecting with him," Martha said. She also noted that she will always be grateful for that "chance encounter" at the solstice picnic and how earnest he was in introducing her to the Piping Plover's habitat.

A Brief Biography of Richard "Dick" English

Born in New Haven in 1935, Dick English was educated locally at the Foote School and the Pomfret School. Then, as a young man, he moved on to Proctor Academy, Nichols College, and the Neighborhood Music School. He earned a Bachelor of Science in business administration from Quinnipiac College in 1961, served in the Army Reserve, and went to work at the First New Haven National Bank, retiring many years later.

He died on July 11, 2011, at the age of seventy-six. Dick never married and had no children. His closest relative was his brother, James. He was active as a friend and mentor to many and a benefactor to his local community. Inspired by his childhood love of symphonic classical music—combined with his own achievements as an accomplished pianist—he became a generous patron of the arts. He supported community groups, such as the Boy Scouts of America and the New Haven Museum. Dick also befriended parks enthusiasts and birders throughout the state of Connecticut.

An Ancestral Legacy of Community Service

The son of Philip and Katharine Dana English, Dick English was a member of a long-standing New Haven family that had called this city home for more than three hundred years. His mother, Katharine, a cellist, was past president of the board of directors of the Neighborhood Music School and the New Haven Colony Historical Society (now the New Haven Museum) and also served as a director of the Young Women's Christian Association (YWCA). Richard's maternal grandfather, Arnold Dana, was involved in preserving Sleeping Giant as part of the Connecticut state park system.

Philip English also had a personal connection to the park system. He served for nearly twenty-five years on New Haven's Parks Commission—which Dick's grandfather, Henry English, had established and served on for fifty-seven years. Henry was also a member of the Proprietors' Committee for the New Haven Green, the Board of Library Commissioners, the Board of Airport Commissioners, and the Board of Education. Dick's paternal grandfather was also a founding member of the New Haven Symphony Orchestra, in 1894.

Other noteworthy family members include Yale University scientist James Dwight Dana, a geologist and chief editor of the *American Journal of Science*. Additionally, Dick was the great-grandson of Connecticut governor James English, who served in that office from 1867 to 1871.

Dick English assumed the mantle of performing good works in keeping with his family heritage and particularly made major contributions to the birding community in the New Haven area—and to all the many bird-lovers who knew him.

Bird walk at Richard English Bird Sanctuary, Killingworth, Connecticut, April 10, 2011. Front row, from left: Deborah Johnson, Gloria Lapolla, Charla Spector, Barney Bontecou, Bobby Lapolla. Rear row, from left: Paul Wolter, Mike Horn, Richard English, Steve Spector, Bill Batsford.
Photo Gina Nichol.

Dick's Love of Birding and the New Haven Bird Club

Those who were close to Dick considered him the "heart and soul" of the NHBC over many decades, and he served as club president in the 1960s. Mike Horn—a close friend of Dick's and coauthor of this tribute—described him as follows: "He had a warm and inclusive spirit and knew more about birds than just about anyone. Yet he was very kind to beginners. He was the one who made me and my wife, Patricia McEnery Horn, feel welcome when we first joined the New Haven Bird Club."

Dick and Mike Horn would later develop and lead an annual early spring bird trip with a stop at the Richard English Bird Sanctuary at the Deer Lake Boy Scout Camp in Killingworth,

Connecticut. Dick was active in the Boy Scouts of America for sixty years and provided significant philanthropic support to the organization. It was in recognition of his contributions to preserving the area's natural habitat that a portion of the acreage at the Deer Lake Boy Scout Camp was designated in his honor as the Richard English Bird Sanctuary.

Dick was also a member of the Connecticut Ornithological Association's Avian Records Committee, the mission of which is to evaluate, document, and tabulate reports of new and rare bird species. He is credited with the first Connecticut sightings of White-winged Dove, Boat-tailed Grackle, Hermit Warbler, and Tropical Kingbird.

Generous in Life, Enduring in Legacy

When Dick English died, it was an enormous blow to his friends and the birding community. In accordance with his wishes—and in line with a family tradition of giving back to the community upon death—the Community Foundation for Greater New Haven (the area's largest grant maker) established an endowment in his name.

At that time, his $20 million bequest was the largest in that foundation's eighty-four-year history. The Richard L. English Fund is now a permanent endowment, invested in perpetuity, from which earnings are distributed as grants annually to organizations that were important to Dick English and his family members. These grant recipients include the New Haven Symphony Orchestra, the New Haven Museum, the Neighborhood Music School, and the Connecticut Yankee Council of the Boy Scouts of America. A separate bequest for $50,000 was made to establish the Richard L. English Fund for Birding Activities in support of the programs of the NHBC.

Mike Horn continues to lead the early spring walk for the NHBC (now renamed the Richard English Memorial Field Trip) and shares fond memories of his close friend with the next generation of birders visiting the Killingworth preserve.

A Soaring Tribute

On Saturday, October 11, 2011, John Triana (the NHBC historian) and Mike Horn led a bird walk to Evergreen Cemetery in New Haven, where Dick English is buried. On a lovely autumn day, the group viewed numerous migrating birds while also visiting the final resting place of some of New Haven's most famous residents.

After an hour or so of birding, the group reached Dick's grave site. Mike Horn remembers what happened next:

> I was then asked to say a few words about him to all assembled. As I was going on and on about our friendship and his birding skills, a big noisy flock of Canada Geese crested the hill and flew in a big V about thirty feet from the top of the English family obelisk. Dick, I don't know how you arranged that signal from the beyond, but a lot of us found religion that day!

References

"Community Foundation for Greater New Haven Receives $20 Million Bequest." *Philanthropy News Digest*, December 19, 2011. https://philanthropynewsdigest.org/news/community-foundation-for-greater-new-haven-receives-20-million-bequest.

McLoughlin, Pamela. "Community Foundation for Greater New Haven Receives Largest Bequest in Its History: $20 Million." *New Haven Register*, December 15, 2011. https://www.nhregister.com/news/article/Community-Foundation-for-Greater-New-Haven-11575666.php.

"NHSO Receives Largest Gift in History." New Haven Symphony Orchestra website, December 16, 2011. https://newhavensymphony.org/nhso-new/

nhso-receives-largest-gift-in-history/.

Richard English, obituary. *New Haven Register*, September 7, 2011. https://www.legacy.com/obituaries/nhregister/obituary.aspx?n=richard-l-english&pid=152558476.

George and Millie
Frank Gallo

The first thing I noticed about George Letis, aside from his warm smile, was his fingers. They were like small sausages—thick, meaty, and round. When he shook my hand, I could feel the calluses from his years of working construction. He introduced his wife, Millie, a jovial woman with kind, intelligent eyes, who reminded me of my grandmother.

George and Millie Letis, February 1983. NHBC archives.

We met in 1980, when I was twenty and studying at Southern Connecticut State College. I was at Long Wharf in New Haven Harbor with a pair of binoculars draped around my neck that I'd borrowed from my ornithology teacher, Dr. Noble Proctor. George and Millie approached and asked me if I was a member of the New Haven Bird Club (NHBC). I wasn't. I'd never even heard of the NHBC. They suggested that I might like to join, given my obvious interest in birds. There was just one problem: the club's meager ten-dollar annual dues were beyond my limited resources. Nevertheless, I was a club member before the day was through. George and Millie had put up the money.

Forces beyond my control were at work. You see, the Letises had a plan. They had been integral to the success of the NHBC for many years. George was president from 1967 to 1969, and Millie was the NHBC's newsletter editor from 1969 to 1989, before becoming club historian in 1989, a position she held into the mid-1990s. During the 1970s, they had realized that the NHBC was quietly dying of old age, and they had taken it upon themselves to remedy the situation.

Before long, I was a NHBC officer, and then I became vice president and, eventually, club president. Following in my footsteps were a host of other young birders recruited by George and Millie—John Himmelman, Tish Noyes, Steve Broker, Andy Brand, Emily Cosenza, and Celia Lewis come to mind, but the list goes on. Each of their stories is remarkably similar to mine, which is not a surprise if you knew George and Millie.

Steve Broker was working at Yale's Peabody Museum of Natural History when he started to get involved in the New Haven birding community in 1980. He recalls:

> Among those who reached out to me in friendship and furthered my learning about birds, Connecticut birding spots, and bird habitat were George and Millie Letis. My situation was not unique. George and Millie welcomed a number of my longtime birding friends to the NHBC, being remarkably gracious to each of us. They knew how to find and make new converts to the wonderful world of birding.
>
> Through the 1980s and later, many of these converts have played key roles in the NHBC. Their names have

been synonymous with the bird club and have included the following, among many others: Noll, English, the Millers, Baratz, the Fletchers, Sibley, Bernard, Siebecker, Rosengren, Borgemeister, the Aimesburys, Massey, Amatruda, Zepko, Shove, Proctor, Schwartz, McBride, Stoddard, Schlesinger, the Lemmons, Cosenza, Zipp, Brand, Mayo, Gallo, Himmelman, Lewis. But the glue that played a central role in moving the club through these years consisted of George and Millie Letis.

Because of the efforts of the Letises, several of us in turn became officers of the club, moving usually from indoor program chair to vice president, then president. During my two-year term as president, I recall a number of board meetings at George and Millie's home on Hunt Lane in East Haven.

In the 1970s, George and Millie also held the Christmas Bird Count (CBC) compilation at their home on Hunt Lane, over which Roger Tory Peterson, aka the "King Penguin" of birding, would occasionally preside. It seems that members of the NHBC had been traveling to Old Lyme, which was Peterson's CBC area, to help him out. George suggested that he return the favor, so Roger started helping with the New Haven CBC. Also at this time, two young upstart birders, Noble Proctor and Davis Finch, were actively birding together. These elements would eventually combine and generate one of George's favorite stories: the day the King Rail ended up in his bathtub.

For several years, Proctor and Finch had claimed to have seen King Rails in their count area; and each year, Peterson would dismiss the sighting, insisting that King Rails did not winter in Connecticut. Noble and Davis grew so frustrated they decided to obtain proof. When the next count day arrived, they found and then managed to capture a King Rail. Unbeknownst to Roger, they took the bird back to the Letises and secreted it in their bathtub. When the time came, they again claimed their King Rail sighting and once more were dismissed by Peterson. Without missing a beat or saying a word, Noble and Davis went to the bathroom, returned with the rail, and placed it snugly in Peterson's lap. King Rail made it on to the list that year.

Since the early 1980s, the Letises house on Hunt Lane in East Haven has been within my CBC area. Faithfully, each year, my team visited the Letises yard during the CBC, once at night to call in their resident Eastern Screech-Owl, and again during the day to record their feeder and garden birds. Their property abutted an extensive tract of undeveloped land, and their yard often hosted uncommon species, including Yellow-bellied Sapsucker, Pileated and Hairy Woodpeckers, and even a Red-bellied Woodpecker (uncommon at the time). There was often a lingering Gray Catbird or Eastern Towhee, Brown Creeper, kinglet, or something else of interest.

The Letises lived at the end of a cul-de-sac, and at night sneaking in and out without waking the neighbors was a bit of a challenge. George's son owned a farm directly across the street, with a flock of domestic geese that were all too willing to sound the alarm when they sensed interlopers. The trick was to go far enough down the driveway to get the screech-owl to respond, but not far enough to alert the geese. We were generally successful . . . except for the times we nearly got arrested.

One memorable night, at 3:30 a.m., having just beaten a hasty retreat from the Letises after waking the geese, we went to the church cemetery around the corner to try again for an Eastern Screech-Owl. As we were pulling back out of the churchyard, a police car coming from the direction of George and Millie's went by us on the main road. I said to my companions: "Watch, he's going to turn around."

I turned out of the driveway and drove in the opposite direction from the cop; sure enough, the police cruiser did a U-turn and came in behind me. He followed us for nearly two miles before I stopped at our next owling spot. He pulled up behind us. I got out with my tape recorder paused on Great Horned Owl.

The officer walked up and asked, "What were you doing in the cemetery?" "Looking for owls," I said. His rather skeptical response was, "In a cemetery?" to which I replied, "Why, you have a better place? Because we didn't find any there."

Surprisingly, we didn't get arrested; the police officer spent the next ten minutes owling with us and was treated to a look at his first Great Horned Owl. I would like to think that the Letises' neighbors were at least partly responsible for piquing an officer's newfound interest in birds. We parted company in good spirits. Subsequently, we have made it a point to inform the police before making our nocturnal sojourns.

Long-standing NHBC member Carol Lemmon recalls doing CBCs with the Letises: "One morning when it was about ten degrees and we were walking in a turnip field in East Haven, Millie went searching for leftover frozen turnips. She said she loved them frozen as they were so sweet then. She located one, cleaned it off on the leg of her jeans, and ate it—around eight a.m. I was so cold, I could not contemplate the thought of eating frozen vegetables."

"Their patience was legendary," she continued. "I saw my first kingfisher with them. They staked out a hole in a bank above a stream, and we were there for nearly an hour before one of the kingfishers flew in the hole."

Tish Noyes notes that George and Millie played an essential role in helping her develop the skills of a birder. "They were two of my favorite people. I spent hours with them at Hammonasset Beach State Park learning birding basics. They encouraged me to pursue it further and study with Noble Proctor. We shared great experiences as well in Costa Rica. While everyone else was off in pursuit of the elusive bellbird, George, Millie, and I stayed at the waterfall in Monteverde watching hundreds of hummingbirds (and listening to a bellbird overhead). I am always reminded of Millie when I see hummingbirds."

On this same trip, I remember standing with Tish and the Letises watching a hummingbird bathing by repeatedly plunging into a rivulet in a stream.

It was well known that George got roped into watching birds by Millie. She was the person more interested in the hobby, and if George wanted to see his wife, he realized he had to take up birding too. Many New Haven birders fondly remember the old station wagon that George had remodeled into a camper of sorts. He'd added a mattress on a platform in the back and mounted a box with their stove and cooking gear on the back door, which swung open. He'd even rigged up a table and awning. George and Millie went on many adventures together across the country in their bird mobile.

Once their sons were old enough to run the construction business, George and Millie always took a month off each summer and traveled around birding. Tish Noyes recalls a wonderful story they told her:

> On one trip out west, they were in an area where three or four states had adjoining borders. They had parked their camper in the parking lot of a laundromat to do some laundry and decided to remain in the lot overnight and continue on their trip at sunrise. The next morning, they left as planned and drove on to an adjacent state. That afternoon, just after entering a third state, they were surrounded by police cars with sirens blaring. It seems that the laundromat two states back had been broken into and robbed the night they slept there. Someone saw

their camper in the parking lot and described it to the police, and an all-points bulletin had been issued to find them. Apparently, they'd been spotted in both the previous states, so the police in all the surrounding states were in hot pursuit! I can just see George, who was nearly deaf, saying to the policemen, "Officers, I didn't hear a thing," and Millie just smiling sweetly.

George had been a contractor, enjoyed tinkering with things, and was always willing to help with any sort of construction project. This was a huge boon to me in my first job as a tern warden/biologist for the Falkner Island Tern Project in Guilford, where I worked with Scott Hopkins, Cindy Pratt, and Dr. Jeff Spendelow. Despite our rather Spartan accommodations—we lived first in an old paint shed and later in the old engine room for the lighthouse—it was one of the best jobs I ever had. We used makeshift furniture, much of which was cobbled together from materials scrounged from the beach, and we slept in hammocks suspended from the ceiling joists. There were no windows, only boards to cover the openings during foul weather—that was until the Letises showed up on the island. George took one look at our living situation and set to work.

He went home to his workshop and returned with real windows and materials for bunk beds, chairs, and tables, which we promptly assembled on site. He even helped us add a wooden floor and a splendid deck using boards from an old dock that had washed ashore. We were living the high life after George was through. Being able to sit outside and watch the sunset on our new deck was wonderful. Unfortunately, the Common Terns essentially saw it as the perfect opportunity to regurgitate fish on us whenever we did.

A much bigger project the Letises involved themselves in was the Smith-Hubbell Wildlife Refuge and Bird Sanctuary at Milford Point. In the 1960s, George was approached by someone who told him that an old hotel on about seven acres at Milford

Point had been deeded to the state as a bird sanctuary. George looked into it and discovered that the "caretaker" living there was basically keeping it as a junkyard (which included a defunct helicopter!). He approached the state and proposed that it lease the property to the NHBC. The state agreed, and George and Millie, with the help of other NHBC members, set to work cleaning it up and making it friendly for birds and birders. They had to remove everything from helicopter parts to junk cars. It took a while, but gradually more and more people began using the sanctuary for its intended purpose. George was manager of the sanctuary in 1968.

Helicopter found at the Smith-Hubbell sanctuary, May 1970, now the Coastal Center at Milford Point. Photo Kevin Gunther.

NHBC member Kevin Gunther was installed as the new caretaker and lived on-site for more than twenty years. The club started having its annual picnic and walks and other events at the sanctuary as well. In the early 1990s, when the Connecticut Audubon Society assumed the lease and built the Coastal Center, George and Millie were there, making sure that the NHBC had a voice in determining the center's future and that the club would maintain a role in assuring that the center would attract and educate future birders.

George and Millie Letis left an extraordinary legacy at NHBC, for which I am deeply grateful. But in addition to being an integral part of the bird club, they also became an integral part of my life.

When George was ninety, I took him on his final birding tour, a three-week trip to South Africa. At the airport when we departed, I ordered a wheelchair to make it easier for him. George was in his glory as an airline attendant moved our entire group to the front of the line and whisked us through security in record time. I can still see him waving his cane at people as we zipped through the airport to our gate.

While visiting the Cape of Good Hope during the tour, we found a photo booth. For a small fee, we could have a photo taken and e-mail it to his family and friends back home. This was in the early days of the internet, and George was thrilled with the idea that he was at one of the ends of the earth and he could send photos to his friends and family.

There was an ice-cream stand beside the photo booth, so afterward, George and I went and bought cones. We were enjoying them and watching the baboons that were sitting around on outcrops overlooking the area, when we noticed a little girl and her father walking hand in hand across the parking lot with their own ice-cream cones. A baboon we'd been watching suddenly jumped off a ledge behind them, raced across the lot, and without missing a stride, grabbed the little girl's ice-cream cone, leaped back up onto the ledge, and quietly sat licking her treat. George and I looked at one another, stunned, then quickly finished our own cones.

In the middle of the tour, we had a half day off in the town of Wakkerstroom, our guide's hometown. I was fighting a cold and really needed a nap. I'd been in my hotel room for about forty-five minutes and had just fallen asleep when there was a knock on my door.

"Who is it?" I asked.

"It's George. I want you to meet some people," came the reply.

I told him I wasn't feeling well and was taking a nap.

His response was classic George: "Come on, let's go. You can sleep when you get home!"

Resigned, I got up and dressed. When I opened the door, George was waiting impatiently in the hallway. As he whisked me toward the street, he proudly explained that he'd been to the baker's next door and had arranged for fresh rice bread to be delivered to the hotel for breakfast (I was gluten-intolerant at the time). George went on to explain that the baker had originally bought the bakery for his son, but the son had decided to leave town. The baker was now stuck with a bakery he didn't want, in a place where he couldn't get wheat with enough gluten to bake good bread.

The man in orange coveralls we'd met at the garage on the way into town was the baker's other son, George informed me as he shepherded me into the pub to meet some of the locals. I was introduced to the bartender, who served us a pint, then was delivered to a table to meet two lovely women dressed in business suits, one of whom, George had learned, was a princess of a local tribe. He went on to explain, with a certain degree of wonder, that "these nice ladies wouldn't have been able to even come into this bar just a few years ago, before the end of apartheid. Imagine that." George seemed to know everyone who walked in the door by name—but then it was just like him to know everyone in town within the first hour of his arrival.

After enjoying their company and meeting a few other locals, we headed out for a walk. We went to see if we could find our guide's house; earlier, he had invited us to stop by. At the end of the street, at the turn to his house, was a fence with a huge expanse of short-grass prairie beyond it. As we stood looking over the grassland, two large birds came flying toward us. They were African Crowned Cranes, birds we had yet to see on the trip, and they winged across the prairie and landed in the tree directly above us.

After a few moments of rapt silence, George reached over and slapped me on the shoulder and said, "See, I told you. You can sleep when you get home!"

Author's Note: Thanks go to Steve Broker, Emily Cosenza, Carol Lemmon, Vanessa Mickan, Tish Noyes, and John Triana for their input and assistance with this tribute.

Dedication of the Hawk Watch Plaque at Lighthouse Point Park

Arne Rosengren

The following is an edited transcription of an impromptu talk New Haven Bird Club (NHBC) member Arne Rosengren delivered on October 8, 1994, at the dedication ceremony of a plaque at Lighthouse Point Park, honoring Ed Shove and the other founders of the Hawk Watch. Florence McBride videotaped the presentation.

Before we had a daily Hawk Watch down here, the New Haven Bird Club used to sponsor a Hawk Watch weekend, usually the third or fourth Saturday of September. I remember very well how George and Millie Letis were very active in that. But there was never a daily Hawk Watch at that time. Members of the bird club and others who were very active birders knew about the place, knew that there were hawks here, and came down to enjoy it.

I'll try to give you an idea of how the daily watch started in 1974. In November of 1971, I joined the bird club, which was a wonderful thing. I've never had any regrets. A wonderful association, not only for the birds but for the friends I've made since joining the club.

> **The New Haven Bird Club**
>
> dedicates this plaque in appreciation of these people of the Lighthouse Point Park Hawk Watch. For over 20 years they gave their time, expertise, and goodwill to help make this the best spot in Connecticut to watch migrating hawks.
>
> We are also grateful to the many others whose participation continues to contribute to the success of this Hawk Watch.
>
> October 8, 1994
>
> Edward Shove Salvator Masotta
> Neil Currie Marjory Pitcher
> Richard English Arne Rosengren
> Millie and George Letis Tony Tortora

Plaque honoring the founders of the Hawk Watch at Lighthouse Point Park, New Haven. Photo Gail Martino.

Following my first club meeting, on a Thursday, I came down on Saturday to bird, and I parked my car. Right over on the open field, there was a fellow standing there with his binoculars and a tripod. I didn't think anything particular of it. I walked around the park for an hour or so, dickey-birding, and when I came back, this fellow was still standing there. Well, curiosity got the better of me. I went up to him and asked him what he was doing. And he said, "Hawk-watching." That was a new phrase to me, an expression I didn't know. This was Neil Currie, by the way. So, I asked if I could join him, and he said sure. We stood there for a couple of hours in mid-November, and we had forty-seven hawks, and half a dozen life birds for me, because I was just starting. And that was the first thing that got me interested in hawk-watching—Neil Currie, down here.

The next two or three years I used to bird every weekend with Dick English, almost every weekend, the year round. We'd go down to Milford Point and come back here in the afternoon. Most of the time on the weekends, Neil would be here. I remember one time asking Neil, "How many hawks have you had?" This was about one o'clock in the afternoon. He said 350. This boggled my mind. Of course, now it's just a routine day, but at the time I'd never heard of 350 hawks in one day.

So that piqued my interest, which led eventually in 1974 to the daily watch. Before the daily watch started there were people who knew hawks, who were hawk-watching sporadically, people like Noble Proctor, and Davis Finch, who used to come down and hawk-watch here and up at the airport on Davis Hill and so on. But not on a daily basis.

Before that, back in the 1950s, there was a fellow named John Cameron Yrizarry, who was a prominent birder down on Long Island and an artist. He was going to Yale at the time, I believe as a graduate student. And he wrote a monograph, which the New Haven Bird Club published and sold for a dollar each at the time, about migration. Not only hawks but everything, spring and fall. He talked about migrations along West Rock Ridge, about East Rock, a lot of places, and he also talked about Lighthouse Point. That was the first specific information that I had ever seen, written information, about Lighthouse Point. He had no specific numbers, but he said there were days when the hawks were just pouring through.

So that's the preface to what happened later with the daily Hawk Watch. In 1974, in September, the New Haven Bird Club had its annual weekend down here. And I was working at the Shubert Theater at the time, and we were not opening until later in the season, in late October. I had four or five weeks off, so

I decided to come down here the following Monday morning, which I did. There was a young girl down here, named Margie Pitcher, who was a senior at Quinnipiac College. She said she was doing the hawk watch to write a paper. And her professor was Dick Bernard, who at that time was president of the bird club. And Dick suggested she come down to do a hawk watch. Margie didn't know anything about hawks, and I knew very little. We teamed up for the first five or six weeks. She could be here three days a week, during the week, and I took the other two days, and sometimes we were down here together.

We wound up the season with sixty-six hundred hawks, and I remember making the announcement at the December meeting, and this huge cheer went up from, oh, probably 150 people at that meeting, up at the Peabody Museum. This cheer went up. I was astonished, but I was also delighted. I was part of what they were cheering about. So, sixty-six hundred seemed like paradise to us at that time.

The next year, Margie had graduated, but she came back, and the two of us were down here daily, for the whole season. We had ten thousand hawks that year. I remember, when we got to ten thousand hawks, Margie and I did a little jig around the area, like a square dance, we were so happy. And all this time, the first two or three years, and ever since, Neil would come down. Neil was teaching at the Taft School in Watertown. He'd have a class that ended at eleven o'clock some mornings, and around twelve o'clock he'd show up here and help us. I remember one time, Margie and I were very embarrassed, because when Neil came down, "Anything going?" he said. "Nah, there's been nothing for the last hour." What we didn't realize was the hawks were going higher and higher, and we weren't looking up. Neil got out of the car and looked up and said, "What's that there?" About thirty hawks were floating around up there. Very embarrassing. But we learned.

Anyway, in '74 we had sixty-six hundred hawks, in '75 we had ten thousand, in '76 we had fourteen thousand, and in '77 we had seventeen thousand, and we've been at or above that ever since. Of course, in the last ten years, it's been above twenty thousand every year. There's a particularly good reason for that, and that's Ed Shove—why the Hawk Watch counts kept going up and up.

Attendees at the Hawk Watch plaque dedication at Lighthouse Point Park, New Haven, 1994. From left to right: Arne Rosengren, Steve Broker, George Letis, Richard "Dick" English, Fred Sibley, Dick Bernard.
Florence McBride collection.

Ed and Richard English and Tony Tortora and three or four others used to come down whenever they could. They would pop in when they had a break and could do it those first three or four years. But other than that, for the first few years, it was Margie and me, and for three or four years after that, I was the only one that was down here during the week. Ed used to come down on weekends.

Then one year—I forget what year it was—there was Ed on Monday morning. I said, "Ed, what are you doing here?" He said, "Well, I'm taking a week of my year's vacation to hawk-watch." That's when I knew Ed was a real hawk-watcher. You don't do that, you don't take a week's vacation just to hawk-watch, unless you're really into it, you really like it. So, then I knew I had a friend.

Then, a few years later, when Ed retired, he started coming down really early, at six in the morning. I never got here until seven-thirty or eight. And Ed started keeping track, keeping the count. After a year or so of that, I suggested to Ed that he become the leader of the count, because he was spending a lot more time down here than I or anyone else was. He didn't want to do that, so finally I said, "Suppose we do it together, be co-leaders, would that be OK?" He said, "That's fine." It was that way for fifteen years, though actually he's been the one who's done all the work. Or most of it.

But anyway, when Ed retired and started coming here every day, that was the luckiest break I ever had for the hawk-watching. It relieved me of the worry of trying to have good coverage down here, of trying to organize it. Ed just quietly assumed the responsibility and took it over. And that was great; it was wonderful to know that even if I couldn't come down or somebody else couldn't come down, Ed was always here.

In the early years, before a lot of people showed up every day, when things were slow, Ed and I talked about this and that, and I never met anybody I respected and liked more than Ed Shove. He was just wonderful. To me it was just a very lucky break that somebody as nice as Ed was the one I was going to spend a great deal of time with down here. And I know we all feel the same way, because I've never heard anybody speak a bad word about Ed—nothing but praise, nothing but friendship and how much they liked him and how much they respected him.

And we now have one of the best hawk watches in the East, and maybe you could say the whole country. We've developed a very fine esprit de corps down here over the years, and everybody's friendly, and everybody enjoys coming down, not only for the hawks but to see their friends every fall when the new season starts and to get together with them.

And that's it in a nutshell, as far as the history of the Hawk Watch goes. I thank you for listening.

The Birdwatchers
Owen Elphick
for my father

There is only one group of people
I can think of
to whom, for instance,
going out into a cold, barren field
at 5 o'clock in the morning
—or perhaps earlier—
and standing there
for the next
several
hours
has any appeal whatsoever,
and that is the birdwatchers,
with their heads tilted back,
their sight soaring into the sky,
shooting out of the barrels of their binoculars,
like bullets
that never hit,
and so never stop;
with their thin notebooks,
where all they need to remember
is recorded;
their thick guide books
where all they need to know
is stored;
and their thicker coats,

where all the heat they need to stay warm,
radiating from their hearts,
which are mindless of the cold,
is trapped;
their sharp, keen eyes
and their keen, sharp ears,
missing nothing of the moment
that stretches around them,
only of the world beyond the field,
beyond reality;
and their studied calm and soft patience,
keeping them in that field
until they decide that perhaps
they should try going
into the woods
to see what they can find there,
hidden from plain sight,
hanging between the world
and the sky,
like them.
There is no group of people
I can think of
who are quite like the birdwatchers,
because the birdwatchers know
exactly what they are looking for,
and, even when they do not,
can always find it
all the same.

Editor's Note: Previously published in the *Hartford Courant*'s Connecticut Poet's Corner (January 2018).

PART VI
Notes from the Field

Bounty Hunter
Frank Mantlik

I got hooked on studying nature, and especially birds, in 1972. Birdwatching, or birding, is a wonderful hobby. It gets one outdoors, communing with nature, in all seasons, in all kinds of weather. Birding can take those who love to travel to places they might never go otherwise. It has gotten into my blood; it's part of my protoplasm. A main focus of my birding in recent years is to discover rare birds. That means being on the lookout for something new, something unexpected. Camera in hand, I'm on the hunt.

What Is a Rare Bird?

There are many possible definitions for a rare bird. The definition most active birders use is: a species that is rarely seen in a given area (such as a state or province) or in a given habitat (you likely won't see a gull in a dense forest) or at a certain, unexpected time of year. A bird unusual or unexpected in an area may not be "rare" in the pure sense.

For example, Wilson's Storm-Petrels are considered fairly rare in the waters of Long Island Sound off Connecticut, but they are an abundant species in their normal breeding range on Antarctic coastlines. In fact, they are considered among the most abundant bird species on the planet. The Dovekie, another pelagic bird, shows a similar pattern. It breeds in enormous numbers in Iceland and Svalbard but in Connecticut is quite rare as a winter visitor. I'm still hoping for one!

So, in a sense, rare birds are like weeds: species out of place.

Of course, some species are truly rare, such as the Spoon-billed Sandpiper, which breeds in Siberia. Its numbers are pathetically low, and the species is in danger of extinction. Other species, such as rails, may not be rare, but they are rarely detected by most birders because of their secretive nature and the dense, grassy marsh habitats they favor. It often takes special effort to get a decent glimpse of one.

One of the attractions of birdwatching is the wonder of flight. Birds have wings with which they can fly across the yard (Northern Cardinal), across the country (Killdeer), or across the globe (Arctic Tern). Occasionally, and for various reasons, individual migrating birds end up off-course. A prime example is the Mistle Thrush (a large European thrush of the genus *Turdus*) that wintered in a neighborhood in New Brunswick, Canada, in 2018. A birding homeowner discovered it in his yard. It is the first record for North America!

Avian Records Committee of Connecticut

The Birds of Connecticut (by John Hall Sage, Louis Bennett Bishop, and Walter Parks Bliss, 1913) and *Connecticut Birds* (by Joseph D. Zeranski and Thomas R. Baptist, 1990) were two authoritative works from different eras that provided an annotated list of the documented species in the state. About twenty-six years ago, since there was no one and no place that kept track of the documented bird species in Connecticut on an ongoing basis, the Connecticut Ornithological Association (COA) created a committee to work on this issue. The Avian Records Committee of Connecticut (ARCC) was born. The committee's principal aim is to provide a complete and accurate record of wild birds reported in Connecticut. A rare records committee can neither verify nor invalidate any records but can provide a judgment on the adequacy of the evidence presented in support of unusual sightings. In other words, this committee, in its rulings, is not saying that a person did or did not see a particular rare bird. Instead, it is ruling on the adequacy of the written documentation and other evidence. The ARCC maintains the official state list (now 445 species) by soliciting reports of rare species and deliberating on these records.

Another repository for documentation of rare species (of birds and other life forms) is the Natural Diversity Data Base. Maintained by the Connecticut Department of Energy and Environmental Protection (DEEP), this database is concerned mainly with state endangered species.

How Does One Find Rare Birds?

Louis Pasteur once said, "In the fields of observation chance favours only the prepared mind." Being open-minded and aware of what is possible greatly increases the chances of a discovery.

In the pursuit of field ornithology and finding rare birds, study the species in field guides, even those that don't normally occur in your area. Observe common species enough so that you become intimately familiar with their appearance, variation, and behavior, to the point that you can confidently identify them with the naked eye. Familiarize yourself with rare species that have occurred in nearby states or regions. This puts otherwise "off-the-wall" potential rare birds on your radar. Look through a flock of common birds and be acutely aware of any that are different in size, shape, color, bill size and shape, behavior, voice, and so on.

Such was the case on October 15, 1985; while driving home from work, I stopped to check out a flock of shorebirds at rain-soaked Veteran's Park in Norwalk, Connecticut. Finding about fifty Pectoral Sandpipers, I glassed through them. One stood out as a bit different: brighter rufous on the cap and breast. I studied this individual and made mental notes of the differences. Arriving home, I raced for my field guides and came to a conclusion: that bird, I thought, was a juvenile Sharp-tailed Sandpiper! This species breeds in Siberia and then is rarely found during migration in North America from Alaska to California.

With darkness falling, I began calling birder friends to describe what I had found. Though I had to work the next day, one friend, Ray Gilbert, was able to relocate the bird and photograph it. Bird bander Dennis Varza even managed to capture it (and several Pectorals), measure it, and release it—removing all doubt that it was anything but a Sharp-tailed Sandpiper (with excellent in-hand photos taken by Robert Winkler). This was the first confirmed record of the species for Connecticut.

Connecticut's first recorded Sharp-tailed Sandpiper (*Calidris acuminata*), discovered October 15, 1985, at Veteran's Park, Norwalk, Connecticut. Photo Robert S. Winkler.

I have had the good fortune subsequently to find other state firsts—Anhinga and Bell's Vireo—as well as several second and third state records. Other state birders with "prepared minds" have also found, and documented, species new to the Connecticut list. The search for rare species is what keeps me going afield day after day. Sure, I never tire of watching the common species, such as an American Robin pulling a worm out of the ground. But I'm hoping someday I'll find a Fieldfare, a (Eurasian) Redwing, or even a Mistle Thrush in Connecticut.

Some people have a knack for finding rare birds. There have been five accepted records in Connecticut of Black Brant, a subspecies of a small goose that breeds in the western North American High Arctic. These have all been documented since 2009—pretty amazing. Even more amazing is that the first four were all discovered and documented by the same person, Nick Bonomo. It's not that other people haven't been searching for one. Couple these Brant sightings with some of Nick's other excellent rare bird discoveries (Red-necked Stint, Mew Gull, Little Egret, Lark Bunting, etc.), and it's clear he has mastered the art of sifting through the chaff of common species for a rare one.

Connecticut has a bevy of experienced bird finders. With more active birders who have more experience and better digital cameras, audio recorders, and other tools, the official list will continue to grow. I find it amazing that at least one new species is confirmed in the state almost every year.

Who Will Believe You? How to Document a Rare Bird

Observe, observe, observe. If you think you may have found a rare bird, make mental or, preferably, written notes of what you see. Draw a sketch, noting field marks. Obtain photographs and/or video. Record vocalizations on your smartphone. It is important to chronicle what you are seeing/hearing *before* you consult field guides. In this way, you minimize any outside influence on what you are actually seeing and noting. Later you can see what your references suggest. Promptly inform friends and/or the general birding community of your discovery, so that others may corroborate and enjoy your rare find.

Communication: Landlines to the Internet Age

WORD OF MOUTH > PHONE CHAIN > RARE BIRD ALERT > PAGERS > LISTSERV > EBIRD

When I began birding in the 1970s, birders in the state were somewhat isolated from each other. The local and regional bird clubs or Audubon chapters would meet once a month. It was then that news of sightings would be shared with friends and co-members. But news was often slow to spread to other clubs or regions, and rarities were often long gone before most learned of them. Birders decided to set up a phone chain, through which news of a rare bird would, theoretically, be passed along from one caller to the next. However, any break in the chain (e.g., someone away on vacation) would slow things down. Then, with the foundation of the COA, a truly statewide organization, the groundwork was laid for more direct communication. The tape-recorded Rare Bird Alert (RBA) was born. Bob Dewire, then I, then other "voices of the RBA" would compile reported bird sightings. Then, once or twice a week, we made an audio recording reporting those sightings on a telephone answering machine. Anyone could call in and find out what birds were around and how to get to them. This method worked fine for about twenty years (and still does today in some cases). Then the British devised an instant alert using a subscription access to electronic pagers. With this technology, new rare finds could be nearly instantly announced to bird-chasing subscribers. Then came computers utilizing the internet, and communication changed in a big way: the Rare Bird Alert message was transcribed to an electronic "page." Through Listserv software, it became possible for anyone to post a sighting directly online, for anyone to instantly see. This is pretty much where we are today. Except now there are also other media and platforms—phone text, Facebook, Twitter, Snapchat, and the like—that people use to communicate instantly.

eBird (www.ebird.org), a project of the Cornell University Lab of Ornithology, is a fantastic bird-sighting database that can be used to submit one's sightings and/or to research the bird data contributed by others. I use it on almost a daily basis, and I encourage others to do so as well.

What's the Big Deal about Rare Birds?

First, it's a thrill to discover a species new to your state or province. I can't even imagine what it must be like to discover a species totally new to science. But if a rare bird has shown up once before, it may show up again.

A prime example is Connecticut's first Pink-footed Goose. Most of us active birders had never even heard of this species when one was discovered among a flock of Canada Geese on a farm in Mansfield, Connecticut, in March 1998. Excellent photos were obtained, notes taken, behavior studied. But at a time when most "rare" waterfowl were assumed to be escaped from captivity (zoos, aviaries, etc.), self-proclaimed "birding geek" Mark Szantyr, a member of the ARCC at the time, was determined to delve into the matter. He contacted zoos, waterfowl breeders, agencies that regulate the same, and even wrote to experts in Europe, where the species occurs. Through his investigations, he learned that the species is almost never kept in captivity and that it was appearing in Iceland and Greenland—leading him to the conclusion that this individual was almost certainly a wild bird that got to Connecticut on its own. The ARCC accepted the record as such. This was a mega-rare bird, representing the first accepted record of Pink-footed Goose in the lower forty-eight states. What's amazing is that, subsequently, the species has become a rare but annual/regular in the northeastern United States and Canada.

What's Next?

It's anyone's guess what bird species new to Connecticut will be discovered next. There are countless possibilities. Surrounding states have documented species that have not yet been found here. For those species that have a pattern of vagrancy to the northeast, it's only a matter of time before Connecticut records them. A prime example is Townsend's Warbler. This western species has been found multiple times in virtually all surrounding states but not yet in the Nutmeg State. One thing is certain: I and many others will be on the hunt!

Author's Note: On April 17, 2020, Paul Desjardins found a Townsend's Warbler at Cedar Hill Cemetery in Hartford. Many got to see and photograph this Connecticut first during its stay of several days. What's next?

References

Natural Diversity Data Base Maps, Connecticut Department of Energy and Environmental Protection. https://portal.ct.gov/DEEP/Endangered-Species/Natural-Diversity-Data-Base-Maps.

Sage, John Hall, Louis Bennett Bishop, and Walter Parks Bliss. *The Birds of Connecticut.* Hartford, CT: State Geological and Natural History Survey, 1913.

Zeranski, Joseph D., and Thomas R. Baptist. *Connecticut Birds.* 1st ed. Lebanon, NH: University Press of New England, 1990.

Through the Lens
Lesley Roy

Ravens on the Cliff

If you had been looking up from the bottom of West Rock Ridge State Park, I would have appeared as a speck perched at the edge of the cliff some three hundred feet from the ground, all to capture photos of the nest site of a special pair of birds.

I was on the lookout for the nest of the Common Ravens that had been seen soaring around the cliff face. You would think a nest that can stretch up to five feet across and two feet deep would be easy to find. When ravens nest on West Rock's distinctive traprock columns, which jut almost vertically skyward, they tend to choose a location under a massive overhang, providing greater protection from predators—and photographers!

Lesley Roy getting ready for her photo shoot at West Rock Ridge State Park in New Haven, 2012. Photo Lesley Roy.

Common Raven (juvenile) perched, 2012. Photo Lesley Roy.

After days of scanning the red-ocher cliff with my binoculars from below, I had finally located the nest, but getting into a position to view and photograph it from the top of the ridge would require substantial effort. A few days later, I set out with a friend to find a good vantage point to photograph the birds. We soon discovered that there was no option but to move farther out onto the cliff ledge than anticipated. We needed to beware of a nearby Peregrine Falcon nest site, also on the cliff face, as these raptors can be territorial if they feel threatened, and they may even attack to defend their young.

Undeterred by these considerations, I carefully climbed down fifteen feet over boulders to perch on a rocky outcropping. I slowly leaned out over the ledge with a tether line attached and found an optimal balancing position. I wedged my left foot between two boulders and stretched out to look straight down the drop of several hundred feet. In a slow, fluid, 180-degree movement, I turned around, motioned to my friend to hand me my camera, pivoted back, and pointed the heavy telephoto lens directly downward. As I then arched my body farther outward, I felt the full force of gravity, and I held my breath, as this was the moment for control. Just then, I began to see the center of the nest and, to my astonishment, one, two, three . . . no, five mottled pale bluish-gray eggs. I took several photos in rapid succession before releasing a deliberate, slow exhale. I got the shot! These pictures serve as valuable documentation for the status of Connecticut's nesting Common Raven population.

Liftoff! Female Peregrine Falcon taking flight from her nest.
Photo Lesley Roy.

Early Inspiration

Growing up as a naturally curious child in the 1960s in rural Connecticut, I was endlessly fascinated to explore the wonderland of vast forests and brooks surrounding my home. Most days, I did not come home until the very last wash of daylight was drawn from the evening sky. Driven by curiosity for creatures both large and small, I made the woodlands my discovery laboratory, as did so many young explorers of this time.

One adventure led to my first encounter with birds. My brother and I heard shouting coming from behind a row of hedges. We ran over only to discover that some young boys had knocked a bird's nest out of a tree and were throwing rocks at it. One boy, close to my age of six or seven, was tossing something up and catching it, and I asked him to show me what he had cupped in his hand. As he reached his arm out, I swiftly scooped up the defenseless, bare-feathered creature and ran as fast as my legs would carry me. The kids were quite upset that I had absconded with their tiny prize. Once parents were involved, it was determined that we put the bird out in the middle of the field and let its parents try to rescue it. I remember the mama bird swooping down frantically; the earlier attack had completely destroyed the nest and killed all the chicks except this one. But this little chick was newly hatched and couldn't even hold its head up. We decided to put it in a shoebox and bring it home. (Now we realize that this was the wrong thing to do; unless you are a rehabilitator, leave the bird where it is!) And so began my love of birds.

At first the chick couldn't open its beak to accept an eyedropper with food, but I kept trying, determined to save its life. Eventually it thrived, survived, and grew into a beautiful blackbird, which we named Tweetie Bird. Tweetie lived with us for a year, roosting in the basement rafters at night. He learned to peck upside down on the kitchen faucet to request a drink.

During the day, Tweetie would come and go through a window over the sink, returning every night. When I was playing in the yard, he would fly down and land on my head, and he always interacted in ways that showed great intelligence and ingenuity. We taught him to forage, lifting rocks to find insects underneath, preparing him to return to the wild. That day came too soon for me, and I cried endlessly but was consoled knowing he was off to live a free and wild life. From time to time we would see Tweetie Bird at our feeders, but he never flew to us again. It was bittersweet, but this wonderful experience laid the foundation for a lifetime of loving winged creatures.

Wild Bird Landing

The most profound bird encounter happened when I was sixteen years old and getting ready to go to school one brisk fall morning. I heard a very loud, croaky bird calling from the top of a tree across the street. In bathrobe and stocking feet, I went out to explore. It wasn't a crow, I was sure—it was too large. I was curious and started to imitate the guttural caw. Instinctually, I put my left arm straight out to invite the bird to come and visit me. To this day I am still surprised that it circled once and swooped right down to land on my wrist. It was a gigantic black bird with shiny feathers, piercing ebony eyes, and talons that dug gently into my skin. As I stood in shock and sheer delight at what was later confirmed to be a Common Raven, I shouted to my mother, "Come quick and bring the camera—something wonderful is happening!" By the time my mom ran out screaming, "Watch out, it's going to poke your eye out," the raven had sidestepped up my arm, and at the moment she snapped a photo, it had its beak in my ear as if whispering secrets!

The next moments seemed like an eternity, as I gently lifted the bird by sliding my hand, with right index finger extended, gently under his talons. As I hoisted the magnificent bird to look into his eyes, my heart was pounding. Then I slowly knelt down and put him on the ground in front of me. My mother was frozen in shock, but I called out to her to take another picture, and she snapped a second magical photo of the bird and me having a conversation. The raven looking up at me, and I back at him, with hands gently outstretched, conveys a story too long to tell here. Those images would end up superimposed over images of my father and his assistant in his laboratory taken the day before; my mom had rewound the film and left it in the camera, accidentally creating melded images of mesmerizing duality, merging the scientific and the natural world together in a miraculous way. The raven's body is superimposed over chemical-filled glass beakers, and my dad in his lab coat appears in ghostly contrast to the winged messenger.

This extraordinary moment has informed my life journey in remarkable and inexplicable ways. I have always followed the path of the raven, and expecting the extraordinary is an empowering mantra repeating throughout my life—no wonder the nickname "Raven Girl" stuck with me!

Photo of Common Raven on Lesley Roy's shoulder superimposed over photo of her father in his laboratory. Photo Lesley Roy.

As a teenager I took up photography, but it wasn't until many years later that photography emerged as a driving force in my quest to document and preserve the environment, its special places, and its precious creatures.

In the spring of 2010, I raised my Canon Rebel starter camera to the sky to photograph birds. With a 100–400mm telephoto lens attached, the whole setup seemed extremely heavy at the time, but now I think of it as my "point and shoot." On that unseasonably warm day on Marginal Drive in West Haven, I heard a squawking ruckus overhead. Looking through the viewfinder, I saw what seemed to be two birds fighting, really having a dustup. It soon became apparent that these birds were not fighting—they were mating!

A slow, quiet approach seemed like natural field craft, and the birds continued flapping wildly until I was in position. Soon I forgot about the tangle of binocular and camera straps, consumed by the task of trying to steady the foot-long lens to hold the focal point locked onto one bird's eye as I pressed the shutter button and blasted off a series of shots.

As the female flew to a nearby tree to rest, that's when I noticed the beautiful colors of the bird's chest, all covered with tiny black, heart-shaped dots—awesome and inspiring.

What were these beautiful creatures? I could barely wait to get home. Once there, plugging the thin memory card into my computer, I experienced something like magic: the most amazing images appeared, ten times larger than life, on the big screen. I shouted out loud, "Are you kidding me?" It turned out that my first bird ID was the eastern, "yellow-shafted" form of the Northern Flicker.

This is the moment you'll be looking for when *you* get behind the lens.

Northern "Yellow-shafted" Flicker, April 2010. Photo Lesley Roy.

Sharing photos with the world to make a difference is nothing new, and the opportunity to reach a vast audience of fellow nature enthusiasts has never been greater. Photography is a great tool to help you identify bird species and educate the next generation. In an ever-changing sea of "glass" (the hip terminology for a photographic lens), technology is breaking the barriers of possibility, allowing photographers to capture what was once unimaginable. Advancements in equipment make entry-level photography easy and fun—light sensors, speed of autofocus, image stabilization, memory card storage, and an endless array of new user-friendly gear make producing great images a snap.

Now, don't get me wrong: it's not as easy as push the button and, *bingo*, you've got an image worthy of a *National Geographic* spread, but the novice or professional photographer has phenomenal advantages not available even five years ago. As you document local birds, habitats, and behavior, and share your images on popular apps like eBird, others will appreciate your birding experience. As greater numbers of people explore photography, the line is blurred between citizen scientist, birder, naturalist, creative artist, teacher, conservationist, and habitat advocate.

Birding through the Lens: The Invisible Becomes Visible

Photographs can reveal details with stunning clarity that you won't otherwise see with the naked eye: the minute spots around a bird's eye ring; tiny feather structures; a streak of cadmium yellow down the wing shaft; the knowing, light-catching glint of a bird's eye. Such images capture much more than field marks: each is a new experience, a singular moment of entry into the world of birds and all that is surprisingly hidden in plain sight.

For me, birding through the lens has nurtured a desire to learn more about the individual behaviors and distinct "personalities" of birds—their territories, favorite food choices, watering holes, mating rituals, nesting, fledging, teaching their chicks, and so much more. Discovery and curiosity fueled my drive to capture the birds' beauty and the endless scope of their habitats, and all the while my skills were improving. Most of what you know technically and have practiced for countless hours and days, if not years, is stored in your body's muscle memory. Oftentimes, I think bird photography should be an Olympic sport! It is challenging, if you want to take it that far. Physical demands can include extreme contortions, endurance, standing for hours craning your neck, lying facedown in the mud, wading in cold waters, sweating, shivering, not to mention the bugs—all in the hope of producing an image that takes your breath away. Yet some images come with ease: while standing on the boardwalk at Wakodahatchee Wetlands in Florida, I got stunning shots of nesting Great Blue Herons six feet from my face and was equally rewarded with the thrill and satisfaction of capturing breathtaking images.

Rewards come in unpredictable ways as well as in epic, heart-pounding moments of profound inspiration and awe—for example, the time I captured two ravens flying in unison, with fully outstretched wing tips touching, a once-in-a-lifetime image. There is nothing more exhilarating than coming across a great birding photo op while out in the field. The rush of adrenaline is instantaneous, as you realize the extraordinary luck and timing that placed you there. Every photographic moment can be a once-in-a-lifetime capture. The good thing about this endeavor is that no matter where you start your photography journey, you will be rewarded richly. Is there any better way to celebrate the world around us than with a photo of a beautiful bird?

Raven pair flying in unison with wing tips touching, August 2012.
Photo Lesley Roy.

Birding photography can expand your horizons, whether you explore the world of birds near your home or on an adventure trip. In 2012, I went to Brazil to photograph the newly discovered Grey-Winged Cotinga. We rented Jeeps and headed up to the highest peaks above the cloud cover. Within minutes, one of the rarest birds on earth flew right past my face with a flash of golden gray. It landed in a nearby tree, where I was able to capture photos later used by Cornell Lab of Ornithology and as a reference photo on the eBird app for South America.

Grey-Winged Cotinga, taken in Brazil, 2012. Photo Lesley Roy.

Some folks travel to birding hot spots to hone their photography skills, while others go to observe the birds and soak up the experience. For me, it is unimaginable not to have a camera along, simply for the sake of recording and documenting. I started supplying images of never-before-photographed birds and others teetering on the brink of extinction, such as Stresemann's Bristlefront (near extinction) and the São Paulo Marsh Antwren. There is opportunity for everyone to wake up their inner explorer, naturalist, and citizen scientist.

Birding documents the fragility of life on this planet and unites us in a global movement to preserve bird species and their nesting and migratory habitats. When I look at my photos, they inspire me to promote new ways of having a gentler impact on all living neighbors.

During the past ten years I have looked through the lens, I have come to embrace the endless and rewarding opportunity for learning. Photography is an entranceway to expand the birding journey. Passing knowledge from expert birders to the younger generation is critical, as interest in the natural world slips away into the virtual world of technology.

Let us be the ones moving the spirit to care for creatures, through mentorship toward stewardship, taking every photographic advantage to educate and inspire "through the lens."

Lesley Roy's Favorite Camera Equipment and Gear

Canon EOS-1DX Mark III DSLR Camera Body with CFexpress next-gen memory cards
Canon EOS 5-D Mark III
Canon 7D Mark II Cropped Body
Canon EF 300mm f/2.8L IS II USM Lens
Canon Extender EF 2X III

Canon EF 100–400mm f/4.5–5.6L IS II USM Lens
Canon EF 600mm f/4.8L IS III USM Lens
Canon EF 24–70mm f/2.8L ii USM Lens
Canon EF 11-24mm f/4L USM Wide-Angle Lens
Canon Speedlite 66EX-RT
Better Beamer Flash Extender
Gitzo GT 4543LS Systematic Series 4 Carbon Fiber Tripod (long)
Wimberley WH-200 Gimbal Tripod Head II or Gitzo Ball Head

Eagle Rock
Deborah Johnson

"So, I'm going to see eagles this weekend; wanna come?" After clarifying that it was not the Eagles *band* I was hoping to see but live birds wintering at a remote spot in Connecticut, this new boyfriend, who was clearly more interested in me than in birds, agreed to come along.

When Bald Eagles started showing up again in Connecticut during the winter in the late 1970s, it was to take advantage of open water where they could find fish. It still took some sleuthing and advance preparation to find them; the World Wide Web, GPS devices, and eBird were active only in the imaginations of scientists and computer geeks (coincidentally, like my new boyfriend). The eagles did some sleuthing of their own and found a ready meal in the water flowing under the ice and over the Shepaug Dam in the Newtown-Sandy Hook area, as fish coming through the sluice are an easy catch in the open water below. I had been to Shepaug only once and had been astounded by the enormous size of the birds and the astonishingly loud crack and echo of the ice breaking off tree limbs as they lifted off for their first morning flight. For this next eagle outing, my plan was to leave New Haven by four o'clock in the morning to arrive at the dam before daybreak—I wanted to share the liftoff spectacle!

"Eagles, really? Are you sure?" he said. "I've never seen an eagle! I thought they were out west, and way up north, not in Connecticut! So, I'm guessing you take this birding thing pretty seriously—it's all new to me, but it sounds like an adventure. A freezing adventure—it's supposed to be well below freezing this weekend. But, sure, I'm game. Where do we go? What time? Shall I drive?"

So off we set in the February predawn, the ground covered many inches deep in snow and ice, and the thermometer registering twelve degrees Farenheight, turning our part of the world into a wonderland of sparkling ice crystals. Despite his offer to drive, we embarked in the low-to-the-ground, two-door Ford Capri I called my "copper bullet," small, lightweight, and trustworthy but already many years old. We arrived, in the dark, at an unmarked access road to the hydroelectric facility on the south bank of the Housatonic River. I edged my little car down a steep incline that ended abruptly at the river's edge. Though I had seen the eagles before from the north side of the river, this south side offered views unobstructed by trees and allowed us to be positioned directly below the birds.

Today there is an eagle observation site on the north side, where River Road dead-ends at the gated entry to the hydroelectric facility. In 1980, long before the enhanced security that followed the 9/11 attacks, there was nothing to stop me or my car. I have wondered in retrospect whether our adventure that day was a factor in the establishment, just a few years later, of the Shepaug Dam Bald Eagle Observation Site. Not only was my car a bit too close to the turbulent waters, I know now that the eagles I found so enthralling are quite sensitive to human intrusion. The monitored eagle observation area is a good thing, for both humans and the birds.

Bald Eagle, Connecticut. Photo Jim Zipp.

Although the Bald Eagle is best known as the national emblem of the United States, the bird's life story is varied and quite special. The species was long classified as endangered on both federal and state levels but has made a comeback, thanks to a number of conservation and management measures. The Connecticut Department of Energy and Environmental Protection (DEEP) provides an excellent background summary of this magnificent bird (see Author's end note).

On that cold morning in February, sunrise was still a half hour off when we got out of the car. It was perfect! Even in the predawn light we could clearly see the dark silhouettes of several eagles. Soon the sun broke through the trees, the branches of which, iced from a recent storm, came ablaze in the morning light against a cold, cloudless winter sky. Now the birds were stirring, and soon the first one lifted off with a silence-shattering crash, then another, and as each lifted, a new shower of ice splintered to the ground, catching the sun in blinding brilliance. The birds soared up and then out over the water, eyeing what the dam had brought them, hungry for their first meal of the day. We watched as at least half a dozen adult Bald Eagles caught breakfast and feasted in the trees or on the frozen river. Young birds—some only a year old, and others perhaps older but still lacking the diagnostic white head and tail of a fully mature bird—begged for food from the more experienced adults. Branches continued to crack in the cold, and birds and ice crystals swirled over our heads. The natural beauty of the scene was ephemeral, soon taking on the appearance of a battle scene, with fish heads, blood, and guts littering the ice on the river. These birds were hungry after a subzero night in the trees and needed the quick energy of a protein breakfast. It was a scene of survival.

Before long, our own survival began to surge into our consciousness as toes and noses began to grow numb. The size and beauty, the pure majesty of the birds was enthralling, and we could have watched them for a much longer time, were it not for the cold. Although I was dressed as warmly as possible with what I owned, that didn't include hand or toe warmers or down-filled garments of any kind. Joseph, no stranger to cold, having grown up in Connecticut and gone to school in Vermont, was wearing a jacket (a slight concession to the cold) but never wore hat or gloves. He would not have suggested calling an end to what I was clearly enjoying—but it was freezing. We admitted to each other that despite relishing our reward for venturing out to find the eagles, we were ready to call it a morning and find somewhere to thaw.

"Shall we go get some coffee and breakfast?" I asked.

"Sure, let's go, but only if you've had enough."

So, in mutual agreement, we slowly turned away from the eagles toward the car and were suddenly blinded by the sunlight glistening on the ice coating the hill that we—or rather I—had driven down in the dark. Our gaze went from the hill, thick with ice, to the lone little car, back to the hill, then to each other, as it struck us both that there may have been some folly in this crack-of-dawn adventure. But anticipating the coffee and warmth that lay somewhere ahead, we got in the car with no thought of not making it up that hill.

On the first try, the tires started spinning about halfway up the hill. Slowly, the car rolled back down to the bottom of the slope, where the river lay. We tried again, and the second try was a repeat of the first. Third time was not a charm. Each slide back down the hill seemed to bring us closer to the river's edge, where the eagles continued feasting, oblivious to our plight. Gunning the car in first gear gave no better result. At this point, we got out of the car and began strategizing. Joseph apologized for not having driven his car, feeling some frustration at not being able to drive a car with a manual shift. The eagles paid us no mind, and as time lapsed and the sun created a slick of melting ice, we paid them less and less attention. Our dilemma had become a test of our new and blooming relationship. Frustrated and cold, we persevered.

"If I drive slightly to the left, I might miss the largest solid sheet of ice," I said. So, I tried that; when the car rolled back down, it was so far to the left that my left rear tire was in the ditch before Joseph hollered, "Stop!" I managed to get all four wheels back on the road, but that diversion into the ditch had unearthed some rocks, and a new strategy was revealed: to add weight to my little car, we began loading into the trunk as many rocks as we could loosen from the grip of the ice. Next try, the car made it two-thirds of the way up, so we knew we were on to something. Back down the hill the car rolled, and the search was on for more rocks, the eagles still noisily calling but now far from our minds.

Joseph was now totally focused on getting us out of there, and adding weight to my little car seemed to be the solution. "If we could free a few more rocks from the ice, it just might do the trick," he said, encouraged.

I was ready to give up, and replied, "Maybe we should just leave it here and come back in the spring."

But Joseph was already using a loose branch to pry up rocks. The ice was thick, the rocks heavy, and the branch splintered. Using his bare, ungloved hands, Joseph picked and scraped and loosened a few more rocks. Another try up the hill: with great anticipation, we neared the top—and hit the patch of ice, by then fully glaring in the sun. Down the car rolled, this time too far to the right and terrifyingly close to the water. But a huge rock stopped it from rolling farther. And with a bit more scraping and pulling with his bare hands, Joseph was able to free that rock and lift it into the trunk. "Eagle Rock," we dubbed it.

After a bit of a pause, Joseph said, "That might do it, but, you know, maybe I should be in the trunk; I weigh almost two hundred pounds . . ."

"You do?! For goodness sake, get out and at least get in the back seat!"

And he did. And the car went up and up and over the top of that hill! Back on the main road, I stopped to let Joseph back into the front seat and to gaze again at the distant trees where I knew the eagles were. Then, with great relief and feeling a bit foolish, yet exhilarated by our exploits, we headed to the Sandy Hook Diner for a huge celebratory breakfast.

We still have Eagle Rock. It came back to New Haven with us in the trunk of the car. That spring it was moved into my garden. Not long after that, it moved to our house. On more than one occasion over the past forty years we've remarked that eagles and Eagle Rock were the rock-solid base of a lasting relationship, because if we could get through that, we could get through anything. In birding as in life, one just never knows what's going to turn up.

Author's Note: Visits to the Shepaug Dam Bald Eagle Observation Site require advance registration but are free; the area is open from mid-December through early March, Tuesday through Friday, and is staffed by volunteers. For reservations call: 800-368-8954.

The Connecticut DEEP's Bald Eagle fact sheet can be accessed at: https://portal.ct.gov/DEEP/Wildlife/Fact-Sheets/Bald-Eagle.

Brainy Birds
Gail Martino

Soon after I moved into my house in New Haven, Connecticut, I started feeding birds on my deck each morning. One day, in addition to placing regular seed in a hanging feeder, I laid three unsalted peanuts, in their shells, on the deck railing. Then I sat by the window, waiting for some activity while sipping a cup of tea.

A single, curious Blue Jay appeared out of nowhere and landed next to the peanuts. No matter how many times I've seen jays do so, I am always amazed at how they recognize peanuts as a potential food source so rapidly. It's as if they are always on the lookout for the largest seedpods on offer.

On this particular morning, as I lingered at the window, what I saw this Blue Jay do next surprised me. The bird picked up the peanut with its pointed black beak and then—instead of flying off with it—returned it to the railing without eating it. Next, he hopped over to the second and then the third peanut and repeated this action. Finally, the bird sidled back over to the first pod, picked it up in his beak, and flew away with it.

"How curious!" I thought. The Blue Jay appeared to be comparing the peanuts, remembering a key difference among them, before selecting the best one. "Is that possible?" I asked myself.

I enjoy pondering this kind of unexpected behavior, because it challenges my assumptions about the cognitive world of birds.

I have always been drawn to questions about how people think, perceive, plan, remember, and problem solve. Wondering about the hows and whys spills over into my hobby as a birder.

Judicious Jays

After I witnessed that Blue Jay's behavior, I decided to see what would happen if I set up three peanuts again. This time, I picked out a longer, three-peanut pod to replace the one the jay had snatched from my deck. Next, I mixed up the order of the peanuts. Not long after, the Blue Jay returned to my deck—but this time it landed immediately next to the longest peanut. Just as quickly, it flew off with this long one in its beak—it didn't compare it to the others, as it had done previously.

Was it by chance, or was the bird actively selecting the biggest seedpod? I added more pods to the deck—some heavier, some lighter—and the bird seemed to always select the biggest or heaviest—at least until another jay arrived. When both jays were feeding on the peanuts simultaneously, they tended to select the one physically closest to their landing position.

"Ah! With a new friend joining the fun, there's no time to be choosy!" I said, laughing out loud.

Before long, this moment of casual observation passed, and I forgot about that morning until, about a decade later, I read an online post about a study by Piotr G. Jablonski, Sang Im Lee, and colleagues from South Korea's Seoul National University. Their findings revealed that Mexican Jays in Arizona both shake and weigh peanuts in their beaks to assess heft as an approximation of nutritional density.

"Blue Jay with a Peanut."
Photo Phillip Turnbull, color modified from original.
Reprinted under Creative Commons license 4.0

Mexican Jays, found in the southwestern part of the United States and Mexico, like the more easterly Blue Jays landing at various times on my deck, are members of the corvid family (Corvidae), which also includes such birds as crows, ravens, magpies, nutcrackers, rooks, jackdaws, and choughs (pronounced *chuffs*). In their study, Jablonski and Lee manipulated the contents of the seedpods, varying the size and number of the nuts inside them, so that they could observe and assess the birds' behaviors. Despite the outer and inner mismatch, the jays reliably selected the heavier peanut by picking up the nuts to gauge their weight. In an interesting twist, the birds were filmed in slow motion for this experiment. Consequently, Jablonski and Lee demonstrated that—even though the human eye couldn't detect it upon normal observation—the birds were rapidly shaking the peanut pods to assess their contents.

The researchers also reported their belief that the birds listened to the resulting rattle and then used the information gleaned (in addition to the weight of the peanut) to help them determine which peanut offered the best meal.

Intelligent Birds?

Observations about bird behavior such as those described above demonstrate something quite interesting to consider: avian cognition—the manner in which birds encode, store, and use information, as well as their ability to perceive, learn, remember, and compare. In the case of the Blue Jays on my deck, the experiment by Lee and colleagues suggests that avian cognition played an important role in how the birds decided which peanut was the best one to snatch for a meal. (For an excellent overview of avian cognition written for laypeople, don't miss Jennifer Ackerman's best-selling book *The Genius of Birds*.)

Among scientists, there is a debate as to whether birds that demonstrate cognitive ability are "intelligent"—a term normally restricted to human cognitive capacity. Charles Darwin suggested, in *Descent of Man*, that animal intelligence differs more in degree than in kind from human intelligence. In other words, he basically viewed intelligence as a measurable attribute with which animals (including humans) were more and less endowed.

Over the years, many nuanced theories of intelligence have been posited. Howard Gardner's seminal book *Frames of Mind: Theory of Multiple Intelligences*, for example, argues for eight different intelligences in humans. One form of evidence for multiple intelligences stems from scientific studies of isolated deficits consequent to brain damage. I saw the implications of this firsthand when I worked at the Aphasia Research Center in Boston in the 1990s, where Gardner held an appointment. On neurological rounds, we would meet patients who, subsequent to a stroke or other neurological damage, could sing but not speak or could speak but not understand the interpersonal nuance of a joke. They were left with "islands" of abilities and disabilities, suggesting different brain areas as responsible for different types of cognition or intelligence.

In recent years, some avian scientists have applied the idea of multiple intelligences to birds—arguing for the existence of capacities such as spatial, vocal, and social intelligences. If this sounds absurd to you, just consider the spatial ability of the Black-capped Chickadee, the bird on the New Haven Bird Club's logo. This small North American songbird, which lives in deciduous woodlands, demonstrates a remarkable skill at recalling the locations of hundreds of stored seeds.

Turns out that the chickadee brain is considered hyperinflated—i.e., heavy for its body size. This hyperinflation is thought to help chickadees store their seeds for safekeeping and remember where they have hidden them. Such brain hyperinflation is also shared by other noticeably clever birds, including jays, crows, magpies, and parrots. If you watch their behavior at a backyard feeder, you are likely to see examples of their cleverness.

Homage to Clever Crows

Some of the best evidence of avian cognitive abilities is displayed by the New Caledonian Crow, a very clever bird found in a small cluster of historically French-controlled islands east of Australia. An endemic species of New Caledonia, measuring sixteen inches of glossy black from head to tail (with a muted, iridescent purple or dark blue sheen when observed in good light), the bird is often referred to locally as the "qua-qua" owing to its distinctive call. Another member of the corvid family, this crow is one of only a handful of birds (and mammals) on the planet that can both use and fabricate tools.

The New Caledonian Crow's varied diet includes eggs, nestlings, small mammals, nuts, seeds, and snails and other invertebrates, particularly grubs. Its "braininess" becomes truly apparent when it hunts for grubs by prodding the invertebrate, tucked inside a cavity, with a stick tool held in its beak. Once the irritated grub bites the stick, the crow withdraws the tool with the grub still attached; then the bird enjoys a nutritious meal. It is thought that its lower beak morphology—a slightly upward angle to the mandible—evolved due to selective pressure to hold a tool straight when hunting in this manner.

Whatever the evolutionary process, the New Caledonian Crow can be seen fabricating its meal-seeking tools by breaking twigs off bushes and trimming them. Moreover, the crows make another sort of tool by tearing strips off the edges of pandanus leaves, which are serrated with barbs that can be used by the crow to hook prey. Further evidence suggests the crows hold the tool so they can see the tip of it. In this way, the crow can aim it, using a preferred eye, as observed by Alex Kacelnik and team at Oxford University's Behavioral Ecology Group.

Not only do New Caledonian Crows manufacture tools, they also appear to invent new tools by modifying existing ones and to pass on these innovations to other individuals in the local group (per Gavin Hunt et al. at the University of Auckland). These incredible crows even invent tools from materials they do not encounter in the wild. In one experiment supporting this finding, such inventive behavior was observed in a crow named Betty, who bent a wire into a hook. This type of evidence suggests that the New Caledonian Crow may be the only nonprimate species capable of cumulative, cultural evolution and generalization in tool use and manufacture (a social intelligence).

Shaken and Stirred by "007"

An episode of the BBC show *Inside the Animal Mind* featured a particular New Caledonian Crow named 007. In this program, the researchers set up an eight-stage puzzle that the bird had to solve, in sequence, to obtain a meat snack. Watching the crow figure out each step is remarkable. First, 007 pulls a string hanging from a perch to release a short stick tied to it. The crow attempts to use this stick to fish out the meat snack, lodged deep inside a small Lucite container. The crow fails and seems to become aware that this stick is too short to serve as a successful tool. Next, in a series of steps, the crow manages to acquire three rocks from different tabletop enclosures with the use of the small stick; he then places the rocks in a bin, which releases a longer tool. Finally, 007 deftly uses the long stick to roll the meat snack backward out of its snug vessel.

Although 007 was familiar with the various items in the puzzle, this crow had never before seen them arranged in that order. This "brainy" bird seemed to deduce the answer to the puzzle step-by-step—as a human might do.

In other experiments, researchers have shown that New Caledonian Crows can use the principle of displacement—Archimedes's law of physics, fundamental to fluid mechanics—to obtain a food reward. A food treat floats on top of a clear cylinder half filled with water, just out of the bird's reach. Like the crow in the Aesop's fable, it then fetches available pebbles and drops them into the water, thereby raising the water level until it can snatch the treat.

When watching footage of such behavior, I am always struck that the crow never overshoots or does more work than necessary. Instead, the bird stops precisely when the water level is just high enough to reach the treat—but no higher.

Due to the New Caledonian Crow's ability to solve such sophisticated cognitive tests, it has become a model species for scientists trying to understand the impact of tool use and manufacture on the evolution of intelligence. In my book, that's one brainy bird!

Sitters and Quitters

The clever feats displayed by the New Caledonian Crow cannot be explained only by an overall larger brain size, although its brain may be large for the bird's weight (as previously described regarding the chickadee). There is more to the story of how such clever birds develop their "braininess," which requires a surprising attribute at birth—complete helplessness.

Some species' early lives begin precariously. Crows, jays, chickadees, and other so-called altricial species begin life in an underdeveloped state, unable to warm or feed themselves; they require considerable parental nurturance to successfully grow to adulthood. Yet, there is an advantage to starting out helpless: these birds experience a longer juvenile period in which they can observe, learn, and store information from their environment and their parents. An extended maturation length is also one of the striking qualities of human development.

It does appear that altricial bird brains—relative to those of precocial birds (which are independent almost from birth)—develop differently in part due to the nurturance of their parents over a longer juvenile period. As Ackerman concludes, nest sitters tend to be brainier than nest quitters.

Start in Your Backyard

You may or may not believe birds like the Blue Jays on my Connecticut deck are truly intelligent. However, it quickly becomes clear to anyone willing to take the time to observe birds that they do display certain degrees of "braininess." And it's not necessary to travel across the globe to a remote island to see this in action. An appreciation for bird smarts can be formed by carefully watching the birds who visit your own backyard.

Observing these living creatures with a curious mind provides clues to the story of their cognitive life.

Consider putting out some feeders, grabbing a cup of tea, and watching your backyard birds. I believe if you observe them intently over time, your answer to the question, "Are birds brainy?" will undoubtedly be "Yes."

References

Ackerman, Jennifer. *The Genius of Birds*. New York: Penguin, 2016.

Commo, Steve. "Betty the Crow Has the Tools to Prove She's the Smartest of Them All." The Independent. February 05, 2014. https://www.independent.co.uk/news/science/betty-the-crow-has-the-tools-to-prove-shes-the-smartest-of-them-all-172650.html

Darwin, Charles. *The Descent of Man, and Selection in Relation to Sex*. London: John Murray. 1871.

Gardner, Howard. *Frames of Mind: Theory of Multiple Intelligences*. New York: Basic Books, 1983.

Hunt, G. R. Tool use by the New Caledonian crow Corvus moneduloides to obtain Cerambycidae from dead wood. Emu 100, 109–114 (2000).

Jablonski, Piotr G., Sang Im Lee, et al. "Proximate Mechanisms of Detecting Nut Properties in a Wild Population of Mexican Jays." (*Aphelocoma ultramarina*). Journal of Ornithology 156 (2015), 163–72. doi:10.1007/s10336-015-1193-6.

Kacelnik, Alex. "Research: Tool Use." Behavioural Ecology Research Group, Department of Zoology, Oxford University. http://users.ox.ac.uk/~kgroup/tools/introduction.shtml.

Matsui, Hiroshi, Gavin Hunt, et al. "Adaptive Bill Morphology for Enhanced Tool Manipulation in New Caledonian Crows." *Scientific Reports* 6(2016), 22776. doi:10.1038/srep22776.

"Mexican Jays Can Detect Nut Properties, Ornithologists Say." *Science News*, May 26, 2015.
"The Problem Solvers." *Inside the Animal Mind,* January 30, 2014. https://www.bbc.co.uk/programmes/p01romd3.
Turnbull, Phillip. Blue Jay with a Peanut. October, 2017. Flickr, https://flic.kr/p/2cnCbYt

Appendices

The New Haven Bird Club and Its Programs

Over many years, the number, range, and popularity of New Haven Bird Club (NHBC) programs have grown tremendously. This would not have been possible without the assistance of a veritable army of dedicated leaders and volunteers, some of whom wrote essays included in this anthology. The following overview is intended to provide a sense of the NHBC's organizational structure and its diverse programs.

Board Composition

The Board of the NHBC (a nonprofit club) functions in support of its programming areas. In conjunction with NHBC board officers (President, Vice President, Treasurer, and Secretary), board members are responsible for delivering the key programs (Outdoor, Indoor, Education, and Conservation). The board members serving as Publicity Chair, Newsletter Chair, Webmaster, and Yearbook Editor play a major role in disseminating information.

The Membership Chair and two members-at-large represent club members' diverse interests at board meetings. Additionally, there is a nominating committee for board appointments. The NHBC also has a subset of members who are not board members but lead specific special programs. One such key member is our historian, who keeps track of NHBC history from year to year and maintains a trove of historical NHBC materials.

Club Programs

Outdoor Programs. At the heart of the NHBC lie the Outdoor Programs, encompassing more than fifty annual bird walks and field trips in Connecticut (and occasionally in bordering states); these outings are led by an experienced NHBC member-leader. While most of these walks occur on weekends, an exception is the monthly First Wednesday Walk. All NHBC bird walks are free and open to the public. They are geared to participants ranging from "newbies" to experienced birders. The NHBC even offers bird walks especially targeted to children and their families.

Indoor Programs. The core of the Indoor Programs is the evening speaker series (now including virtual meetings). This takes place after the monthly NHBC meeting (held from September to May). A speaker from the birding community is allotted an hour for a presentation to attendees. Speakers are drawn from within and outside Connecticut to promote idea cross-fertilization. Free and open to the public, the speaker series covers a wide variety of topics. Past appearances have included a lecture on the science of migration, a photo presentation from a professional nature photographer's portfolio, live "avian ambassadors," a falconer with a live falcon, and scientist Richard Prum on the evolution of beauty and mate choice. The Indoor Programs are occasionally held off-site. One such event was a tour of the bird skin and nest collections at Yale's Peabody Museum of Natural History (managed by Drs. Kristof Zyskowski and Richard Prum). For more than two decades, the NHBC has been fortunate to have member Stacy Hanks prepare themed refreshments for the enjoyment of Indoor Programs attendees.

Education. The NHBC is involved in local public schools and camps to encourage children to participate in birding. A basic birding skills class geared to children is followed by an hour-long bird walk where the children can practice using binoculars. To build on relationships with young people, the Education program annually spearheads scholarships and/or grants for students to learn more about birds and nature in school and camp programs. The program also hosts events for adults and families. Leaders of the Education program represent the NHBC at events such as the Connecticut Green Expo and the Lighthouse Point Park Migration Festival.

Conservation. The NHBC works to preserve and restore bird populations in their habitats, as well as to provide opportunities for learning about birds, habitats, and ways to ensure their future environmentally. The Conservation Program's activities include beach cleanup, planting of bird-attracting flora, community education about creating bird-friendly habitats, and environmentally focused, family-friendly walks. The program has developed an ecosystem of collaborating organizations with goals in alignment with those of the NHBC, partnering with the City of New Haven, local land trusts, Friends of East Rock Park, and Friends of Stewart B. McKinney National Wildlife Refuge, among others.

Special Programs and Events

In addition to its regular programs, the NHBC participates in special programs, events, and activities at the international, national, state, regional, and local levels—some led by the club, some co-led with other organizations.

- Christmas Bird Count (December). The longest-standing birding event in which the NHBC participates and the longest-running citizen science project in the world, the Christmas Bird Count is an annual survey of winter birds. In the United States, the results of the count from each location are sent to the National Audubon Society for inclusion in the international census of early winter bird populations.

- The Big Sit!® (October). This annual event, created by NHBC member John Himmelman in 1992, has become an international bird count with participants in over two dozen countries. The objective of this event is to record as many species of birds as possible within twenty-four hours from a seventeen-foot-diameter circle. Participants can join other birders in one of the established circles or create their own circle. *Bird Watcher's Digest* has organized and coordinated the event since 2001, but this responsibility reverted to the NHBC in 2020.

- Big Day Birding Marathon (May). Like many bird clubs, the NHBC organizes a one-day birding marathon during the height of spring migration. The information is logged into eBird and compiled locally, regionally, and internationally.

- Connecticut Bird Atlas (2018-2021). The NHBC is participating in the Connecticut Bird Atlas Project, a joint effort of the Connecticut Department of Energy and Environmental Protection and the University of Connecticut. The goal of the project is to map all species found in the state during both breeding and nonbreeding seasons.

- Hawk Watch (September–December). Led by a committed team of hawk-watchers, this is a long NHBC tradition. Each September to December, the team scans the skies from Lighthouse Point Park in New Haven, recording the migrating raptors and other species sighted.

- Mega Bowl of Birding (February). This event was inspired by the Super Bowl of Birding conducted each January by Mass Audubon. Held on NFL Super Bowl weekend, the Mega Bowl is a twelve-hour competition in which teams of birders of all ages and abilities attempt to find the greatest number of species and amass the most points, which are based on the rarity of the birds recorded.
- Summer Bird Count (June). The goal of this weekend-long event is to identify as many species as possible in one's local area. Unlike other counts, which emphasize sightings of birds, this offers a great opportunity to develop skills in song identification, because many of the birds are entirely hidden in the summer foliage.
- Migration Festival (September). An annual New Haven Parks Department program, held at Lighthouse Point Park, during which the NHBC typically offers activities such as bird walks, guided hawk-watching, and an educational table.
- Winter Feeder Survey (Winter). The Winter Feeder Survey is a yearly census to determine the number and frequency of birds visiting feeders in the greater New Haven area. Members are invited to watch and record the activity at backyard bird feeders at least once a week.
- Rare Bird Sightings (ongoing). NHBC members often post rare and noteworthy bird sightings at the American Birding Association's birding news lists (see: http://birding.aba.org/mobiledigest/CT01).
- Binocular Donation (ongoing). The NHBC collects binoculars, scopes, and neotropical field guides that are then donated through the American Birding Association's

Birders' Exchange Program to birders in Latin America and the Caribbean.
- Birds in Words Book Club (bimonthly). NHBC-hosted book discussion group.
- Book Sale (November). The NHBC's annual sale of donated birding and nature books is normally held at the monthly Indoor Program meeting in November.
- Ride-Sharing Program (ongoing). The NHBC ride-sharing program enables birders lacking access to transportation to attend in-person NHBC events.

Publicity

Publicity for the NHBC and programs can be found on these platforms:
- NHBC Website (www.newhavenbirdclub.org). This is the first place to check for upcoming meetings, programs, and other information.
- Facebook, Instagram. The NHBC can be found on leading social media sites.
- *The Chickadee*. The bimonthly NHBC member newsletter summarizes all upcoming activities. Past and current issues can be found at the NHBC website.
- NHBC Yearbook. The members-only annual booklet details the upcoming program-year events and member contact information.

Appendices

Twenty-Five Fun Facts about the New Haven Bird Club

1. The New Haven Bird Club (NHBC) is one of the oldest bird clubs in the United States. The name "New Haven Bird Club" first appeared in the publication *Bird Lore* in an account of the 1906 Christmas Bird Count. This mention predated the club's formal establishment.

2. The NHBC was formally established on April 3, 1907, by a group that included a number of New Haven High School and Yale University students.

3. Katherine Urban was the first noted historian for the NHBC, beginning in the 1968–69 club year. Today our club historian, John Triana, maintains more than 113 years of records, publications, and history.

4. The NHBC was established around the same time as the National Audubon Society of the United States.

5. The NHBC is one of approximately five hundred bird clubs currently active in North America.

6. The NHBC has had fifty-one Presidents since inception.

7. The NHBC has always welcomed women as members and in leadership positions. There have been sixteen women Presidents of the NHBC to date.

8. From the very beginning, the club identified the importance of conservation. One of the first conservation efforts consisted of women members lobbying local merchants to discontinue selling hats decorated with aigrettes made from breeding bird feathers.

9. The goals of the NHBC are to create opportunities for recreation and education in birdwatching and to participate in the conservation of natural resources in New Haven and surrounding areas.

10. The NHBC hosts a variety of activities to address many levels of interest, including bird walks, a speaker series, citizen science bird counts, conservation activities, children's educational activities, and even a book club!

11. The NHBC leads about fifty bird walks each year.

12. The New Haven area is productive for birdwatching across all four seasons. Some favorite spots for NHBC bird walks include East Rock Park (spring), Lighthouse Point Park (fall), and Sandy Point Beach and Bird Sanctuary (winter and summer).

13. Within and around New Haven, the National Audubon Society has designated five Important Bird Areas for the Atlantic Flyway. Two are globally important: Sandy Point in West Haven and the Quinnipiac River Tidal Marsh. NHBC members participate in conservation efforts to preserve these and many other locations around Connecticut.

Appendices

14. The NHBC has many ties to Yale University, although there is no formal affiliation. Some NHBC founders were Yale students, and in years past, the club's meeting place was often on the university campus.

15. One former NHBC President was Yale University ornithologist Fred Sibley, father of David Sibley, creator of the popular *Sibley Guide to Birds*. One of the oldest members, Arne Rosengren (in his mid-nineties as of this writing), fondly remembers a young David drawing shorebirds at Sandy Point in West Haven while accompanying his father. David is an honorary life member of the NHBC.

16. The club logo is a perching Black-capped Chickadee encircled by the words "New Haven Bird Club, established 1907."

17. Current membership dues are fifteen dollars a year per individual membership. Corrected for inflation, this is not much more than dues (fifty cents annually) in 1907, when the club was established.

18. The NHBC's newsletter, *The Chickadee*, appears six times a year. The club has published an annual yearbook of upcoming events since 1917, skipping only a handful of years.

19. The NHBC publishes a checklist of the birds of New Haven County, which can be obtained by contacting the club via the website.

20. The NHBC has many longtime members—some have been in the club for over forty-five years! They came for the birds and stayed for the friendships and connections they made.

21. The NHBC is one of several birding organizations in Connecticut; others include Audubon Connecticut, Connecticut Audubon Society, Connecticut Ornithological Association, Connecticut Young Birders Club, and Hartford Audubon Society. Many NHBC members are also members of some of these other associations, allowing for cross-fertilization of ideas, educational activities, and event co-sponsorship.

22. Former NHBC member Noble Proctor, who passed away in 2015, is still credited with the largest number of species identified in Connecticut, at 404, according to eBird.

23. Club member John C. Himmelman created The Big Sit!® bird count, now an international birding event promoted by *Bird Watcher's Digest*, which coordinated the event and brought it to a wider audience through an internet interface. In 2020, the NHBC resumed active management of The Big Sit!

24. The advocacy of NHBC member Ernest F. Coe was instrumental in the establishment of Everglades National Park, dedicated in 1947.

25. The NHBC Hawk Watch at Lighthouse Point Park logs between six thousand and eight thousand raptors each fall migration. Among the species sighted are Cooper's Hawk, Sharp-shinned Hawk, Bald Eagle, American Kestrel, Northern Harrier, Merlin, Osprey, Peregrine Falcon, and even an occasional Golden Eagle. Although that sounds like a lot of birds, in the early 1990s, Hawk Watch totals were over twenty thousand each fall; habitat loss, insecticides/rodenticides, and environmental changes affect the lives and breeding of these birds.

About the Authors

DeWitt Allen joined the New Haven Bird Club immediately on moving to Hamden in 2012. He has served on the NHBC Board as Member-at-Large, Membership Chair, Vice President, and President.

Martha Lee Asarisi, born, raised, educated, and employed in Connecticut, is a longtime pharmacist and a member of the University of Connecticut School of Pharmacy Alumni Board. An Osprey steward for Connecticut Audubon, she has been a birder for more than thirty years and member of NHBC since 2008. She and her husband (fellow club member Richard) and daughter Natalie reside in Bethany, Connecticut.

Dan Barvir served as a ranger for the New Haven Parks Department for thirty-five years.

Ricci Cummings found her way to birding and the New Haven Bird Club through marriage. A former clinical social worker, Ricci is a retired corporate litigation attorney. She is President of the Board of Clifford Beers, a board member of Mid-Fairfield Child Guidance Center, a commissioner on the Hamden Zoning Board of Appeals, and on the advisory board of the Center for Ethnic, Racial, and Religious Understanding at the City University of New York. She is the proud Bubbie of five glorious grandchildren.

OWEN ELPHICK, a writer and performer from Storrs, Connecticut, is the son of two University of Connecticut ornithologists and grew up around a community of birders—an experience that informs his poem "The Birdwatchers" as well as several other works. Owen was a 2015 Poetry Out Loud state champion and national finalist, and a 2016 winner of the Hill-Stead Museum's Fresh Voices Poetry Competition. His work has been honored by the Connecticut Drama Association, the National Collegiate Honors Conference, and several other organizations. His first book, *Thoughts & Prayers*, was published in 2019. He has a BFA in creative writing from Emerson College.

FRANK GALLO is a long-standing member of the NHBC and served as club President from 1989 to 1991. He recently published *Birding in Connecticut* (2018) and has published two children's books, *Bird Calls* and *Night Sounds*. He is also a professional bird tour leader for Sunrise Birding, LLC, and is former director of the Connecticut Audubon Society's Coastal Center at Milford Point.

MICHAEL HORN joined the New Haven Bird Club in 1993 and has served in various positions, including President (2011–13); he is currently a life member. He is Vice President of Whitman Controls in Bristol, Connecticut, and has been chief judge for the annual New Haven Science Fair for twenty-one years. A close friend of Richard English, Mike now serves as walk leader for the Richard English Memorial Field Trip, from Lighthouse Point to the Richard English Bird Sanctuary in Killingworth, an all-day event that takes place in early spring.

DEBORAH JOHNSON is a life member of the NHBC. She currently serves on the board as Conservation Chair.

PAT LEAHY has been the "landlord" of at least seventy Eastern Bluebird/Tree Swallow nest boxes on Regional Water Authority property in Bethany and Woodbridge, Connecticut, for the past fifteen years. He has been on the Board of the NHBC for more than twenty-five years, with a few breaks, and is currently the Webmaster. Retired from an IT career, Pat is the chef at his church's once-a-week soup kitchen in downtown New Haven, serving three hundred to five hundred meals each Wednesday.

CHRIS LOSCALZO is the Mega Bowl founder and coordinator. A member of the NHBC for thirty years, he has served as President and the Outdoor Programs Chair. He has been the compiler for the New Haven Christmas Bird Count for twenty-five years, is the current President of the Connecticut Ornithological Association, and has had an article on birding published in *Bird Watcher's Digest*. Chris planted and maintains a large native shrub garden and fruit tree orchard in the Woodbridge Community Gardens. He is a clinical cardiologist at the Yale Heart and Vascular Center.

PATRICK LYNCH is an artist and author, and lives in North Haven, Connecticut.

FRANK MANTLIK, an avid birder, naturalist, and photographer for more than forty years, has been active in many bird and nature organizations in New England and has served as President of the Connecticut Ornithological Association. He leads several walks for the NHBC and is a professional birding tour guide for Sunrise Birding, LLC.

GAIL MARTINO, PhD, has been a member of the NHBC since 2009 and now chairs the club's Indoor Programs. For the past three years she has served as a surveyor for the Connecticut Bird Atlas project, whose goal is to document the distribution and abundance of Connecticut bird species during breeding and nonbreeding periods. Originally trained as a neuroscientist, she has held several academic and corporate positions, and is currently a Senior Manager of Innovation at Unilever. In 2020, Gail published her first children's book, *A Friend for Bloo*, an adventure story about a bird looking for friends in the forest.

STEVE MAYO has been a member of the NHBC and a visitor to Lighthouse Point Park Hawk Watch since 1986. He has been the Hawk Watch compiler since 2007. Steve has served as Field Trip Chair and Vice President of the NHBC. He is a past President of the Connecticut Ornithological Association and is a NorthEast Hawk Watch Association board member.

FLORENCE MCBRIDE is a long-standing NHBC member and educator, who leads walks for both children and adults. She was named the national Nature Educator of the Year (School-based) by the Roger Tory Peterson Institute in 1993 and has received a Lifetime Achievement Award from the Connecticut Science Teachers Association, the Connecticut Ornithological Association's Mabel Osgood Wright Award, and the New Haven Bird Club's President's Award for outstanding service to the club. She is committed to sharing her educational materials with others who want to help children observe and learn about birds.

CRAIG REPASZ, an NHBC member since the early 2000s, has served on the Board as Indoor Programs Chair, Conservation Chair, Vice President, and President, and is currently the Outdoor Programs Chair. He holds a BA in the history of science and medicine and has published articles on medical sectarianism and on changing disease conceptualization. Now retired, Craig is the Connecticut Bird Atlas volunteer coordinator, President of the Friends of Stewart B. McKinney National Wildlife Refuge, an alpine steward in the White Mountains of New Hampshire, and board member of the Connecticut Ornithological Association and the Hamden Land Conservation Trust.

ARNE ROSENGREN has been active in the NHBC since 1971. He was instrumental in founding the daily Hawk Watch at Lighthouse Point Park and served as President of the club from 1977 to 1979. Since then, he has been one of the most recognizable and popular figures in the bird club, at meetings and on field trips around Connecticut.

LESLEY ROY is a professional photographer and journalist whose photos have been published in *New Haven Magazine* and *National Geographic* and have been featured by the National Audubon Society, Connecticut Audubon, Connecticut Ornithological Association, eBird South America, and eBird Central America. She is a lifetime member of the NHBC. As a City of New Haven Commissioner for Arts, Culture, and Tourism (2014–19), Lesley published *Soaring in New Haven*, a photography book celebrating the region's birds and habitats, which is presented as a gift to foreign officials, dignitaries, and other visitors. She is the owner of Lesley Roy Home Couture (www.LesleyRoy.com). Her photography website is QuantumWow.com.

TOM SAYERS is the founder of the Northeast Connecticut Kestrel Project. Information on this project can be found at nectkestrels.com.

Acknowledgments

There are many people to thank for helping this project take flight. First is Craig Repasz, former President of the New Haven Bird Club (NHBC). He recognized the value of capturing member stories, saw the potential in this project, and was an early supporter. His essay, "A Century and More," provides a valuable overview of NHBC history. DeWitt Allen, current club President, continued the support by serving on the book committee as executive member and helping us obtain the resources we needed to bring the project over the finish line. We owe him a world of thanks.

Additionally, NHBC board members are acknowledged for getting behind the project: Bill Batsford, Donna Batsford, Lori Datlow, Julie Hart, Christine Howe, Genevieve Nuttall, Laurie Reynolds, Andy Stack, Charles Strasser, and Pete Vitali. Indeed, some board members contributed to or wrote essays: Mike Horn, Deborah Johnson, Patrick Leahy, and Alan Malina.

A special thank you goes out to the other contributors (not part of the Board) who volunteered their time in the midst of their busy schedules to capture the stories contained herein: Martha Lee Asarisi, Dan Barvir, Ricci Cummings, Owen Elphick, Frank Gallo, Chris Loscalzo, Patrick Lynch, Frank Mantlik, Steve Mayo, Florence McBride, Arne Rosengren, Lesley Roy, and Tom Sayers. Added to this list are the many members who contributed stories and anecdotes or sat for interviews that became parts of essays: Gilles Carter, Glen Cummings, Tina Green, Valerie Milewski, Beverly Propen, and Jack Swatt. A key member who assisted with content across various pieces is our club historian, John Triana.

Many additional club members contributed thoughts, comments, and suggestions about content in this book as part of their support of the club's Indoor Programs. Their contributions were most welcome and valued.

There are many, many other members of the club whose names may not appear in these pages but whose contributions to the club's success are no less vital and appreciated. It's for this reason we dedicate the book to all these members, past and present.

We also acknowledge Jim Zipp, NHBC member and professional photographer, for allowing us to use his photo of a chickadee on the front cover. A life member, he has served as speaker for the Indoor Programs several times, sharing his outstanding photographic collections of birds and mammals. More of his incredible work can be found at JimZipp.com

We thank those who provided photos for use in the book: Bill Batsford, Dan Cinotti, Christine Howe, Florence McBride, Gina Nichol, Northeast Connecticut Kestrel Project, Lesley Roy, Abby Sesselberg, John Triana, Phillip Turnbull, Peter Whiteford, Robert S. Winkler, and Jim Zipp.

There are members of the production team without whom this book may never have been published. This includes our editor, Amy K. Hughes, who was such a pleasure to work with. Her experience, professionalism, and perspective helped create the book's final shape. Our book designer, Hannah Gaskamp, was incredibly helpful in crafting the design of the cover and the interior layout. Thanks also to our proofreaders: Donna and Bill Batsford, DeWitt Allen, and Ricci Cummings.

Last, but certainly not least, I must thank my co-compiler, Ricci Cummings. I am so grateful that she answered a call for a volunteer to help me with this project. Ricci selflessly gave her time to the effort. Trained as a lawyer, she applied her sharp mind to help shape many of the essays herein. Having never met before, Ricci and I became good friends over the course of this project. As with many large, lengthy projects with many moving parts, there were some ups and downs along the way. Ricci's support never wavered. Her positive energy never subsided. It has been a special gift in my life to meet, work, and bird with her.

—Gail Martino

Made in the USA
Middletown, DE
18 May 2021